PEGGY ORENSTEIN

AUTHOR OF *CINDERELLA ATE MY DAUGHTER*

GIRLS & SEX

NAVIGATING THE COMPLICATED NEW LANDSCAPE

PRAISE FOR
Girls & Sex

"Thought-provoking. . . . The interesting question at the heart of *Girls and Sex* is not really whether things are better or worse for girls. It's why—at a time when women graduate from college at higher rates than men and are closing the wage gap—aren't young women more satisfied with their most intimate relationships? . . . *Girls and Sex* is full of thoughtful concern and empathethic questions."

—Cindi Leive, *New York Times Book Review*, cover review

"A nuanced read for anyone who remembers being a young woman and anyone who is raising the next generation of girls (and boys) for whom we hope the future holds sexual satisfaction, not pain or disappointment." —Rebecca Traister, *More*

"Provocative and thoughtful. . . . Both an examination of sexual culture and a guide on how to improve it. . . . The breadth of Orenstein's reporting . . . is impressive."

—Laura Stepp, *Washington Post*

"An honest and thoughtful exploration. . . . It would be easy to pigeonhole *Girls & Sex* as essential reading only for parents of female teens or preteens. . . . [But] this book is for anyone who cares for a girl approaching womanhood."

—Adrian Liang, Amazon Best Book of the Month citation

"*Girls & Sex* should be mandatory for anyone who cares about the present and future cultural landscape for girls, women, humans. I seriously want to quit my job and tour the country, furiously hawking Peggy Orenstein's insightful, important book."

—Rashida Jones, actress, writer, and producer

"[An] important new book. . . . Her writing is clear and compelling, her analysis is incisive and thorough, and her findings are downright troubling." —Sharon Holbrook, *Washington Post*

"Nonsensational but deeply entertaining. . . . A must-read." —*People*, Book of the Week

"Fascinating. . . . A wise and sharply argued look at how girls are navigating 'the complicated new landscape' of sex and sexuality." —*The Economist*

"An intimate view of the sex lives of young women in the United States. While revealing disturbing common threads . . . Orenstein brings levity to this fraught topic." —*Elle*

"*Girls & Sex* may do more to change how sex education is rethought and how parents and daughters discuss pleasure and sexuality than any book since the landmark *Our Bodies, Ourselves*." —Tamara Straus, *San Francisco Chronicle*

"If you're going to talk about women in the twenty-first century, you MUST read Peggy Orenstein's *Girls & Sex*. No one else is asking these questions; so no one else, then, is finding out the answers." —Caitlin Moran, author of *How to Be a Woman*

"A smart, sobering guide to the sexual lives of young women today." —Ann Levin, Associated Press

"I'm not going to tell you to go right now and buy a copy. . . . I'm going to tell you to buy two copies: One for yourself, and one for the teenager in your life. . . . Refuses to be judgmental or doom and gloom. Instead, it offers something else—a demand for education, enlightenment, and ultimately, the radical notion of equal satisfaction." —Mary Elizabeth Williams, *Salon*

ents and young girls the power to make informed decisions regarding sex."   —*Kirkus Reviews*

"Sex and teenagers have always gone together, but parents reading Orenstein's frank exploration of current trends may still be in for a shock. . . . This isn't a comfortable book to read (Orenstein herself admits twinges a few times), but it's an important one."
                                                                —*Booklist*

"Accessible prose and narrative style will bring the work of many thoughtful experts to a wider audience. . . . Young adults, parents, educators, and activists alike will find this passionate work a timely conversation starter."   —*Library Journal*

GIRLS
& SEX

ALSO BY PEGGY ORENSTEIN

Cinderella Ate My Daughter: Dispatches from the Front Lines of the New Girlie-Girl Culture

Waiting for Daisy: A Tale of Two Continents, Three Religions, Five Infertility Doctors, an Oscar, an Atomic Bomb, a Romantic Night, and One Woman's Quest to Become a Mother

Flux: Women on Sex, Work, Love, Kids, and Life in a Half-Changed World

Schoolgirls: Young Women, Self-Esteem, and the Confidence Gap

PEGGY ORENSTEIN

GIRLS & SEX

NAVIGATING THE COMPLICATED
NEW LANDSCAPE

HARPER

NEW YORK · LONDON · TORONTO · SYDNEY

A hardcover edition of this book was published in 2016 by HarperCollins Publishers.

HarperCollins books may be purchased for educational, business, or sales promotional use. For information, please e-mail the Special Markets Department at SPsales@harpercollins.com.

FIRST HARPER PAPERBACK EDITION PUBLISHED 2017.

Designed by Lisa Stokes

Library of Congress Cataloging-in-Publication Data has been applied for.

ISBN 978-0-06-220974-0 (pbk.)

17 18 19 20 21   ov/lsc   10 9 8 7 6 5 4 3 2 1

For my one daughter, my eight nieces, my two nephews,
and all the girls and boys I've met along the way

Contents

Contents

GIRLS
& SEX

Everything You Never Wanted to Know About Girls and Sex (but Really Need to Ask)

A few years ago I realized that my daughter wouldn't be a little girl much longer. She was headed toward adolescence, and honestly, it put me in a bit of a panic. Way back in preschool, when she was swanning around in her Cinderella gown, I took a deep dive into the princess industrial complex and came back convinced that its seemingly innocent pink-and-pretty culture was priming little girls for something more insidious later on. Well, "later on" was now coming at us like a Mack truck—a Mack truck whose driver was wearing five-inch heels and a micro-mini, and was checking her Instagram when she ought to have been looking at the road. I'd heard horror stories from friends with teenagers about how girls were treated in the so-called hookup culture; of girls coerced into sexting or victimized in social media scandals; of omnipresent porn.

I was supposed to be the expert at decoding the mixed messages of girlhood. I traveled the country schooling parents on the difference between sexualization and sexuality. "When little girls play at 'sexy' before they even understand the word," I'd tell them, "they learn that sex is a performance rather than a felt experience." True enough. But what about once they *did* understand the word?

It wasn't as if I had any answers. I, too, was just trying my best to raise a healthy daughter at a time when celebrities presented self-objectification as a source of strength, power, and independence; when looking desirable seemed a substitute for feeling desire; when *Fifty Shades of Grey*, with its neurasthenic lip-chewing heroine and creepy stalker billionaire, was being hailed as the ultimate feminine fantasy; when no woman under the age of forty appeared to have pubic hair. Sure, as a girl I wore out songs such as "Sexual Healing" and "Like a Virgin," but they were Disney Channel fodder compared to L'il Wayne's "bitch" whose "strict diet" in the song "Love Me" consists of nothing but "dick"; or Maroon 5's promise to hunt a woman down and eat her alive in "Animals." (In the video, lead singer Adam Levine stalks the object of his obsession while dressed as a butcher wielding a meat hook, then has sex with her in a blood-drenched finale.) It's enough to make me apologize to Tipper Gore for the way my friends and I mocked her in the '90s. Meanwhile, study after study has revealed a shocking prevalence of sexual assault on college campuses; the problem is so dire that the president of the United States (himself the father of two teen girls) has become involved.

Even as girls outnumbered boys in college, as they "leaned in" to achieve their academic and professional dreams, I had to wonder: Were we moving forward or backward? Did today's young women have more freedom than their mothers to shape their sexual encounters, more influence and more control within them? Were they better able to resist stigma, better equipped to explore joy? And if not, why not? Girls now live in a culture where, increasingly, unless both parties agree unequivocally to a sexual encounter, there is no consent—only "yes means yes." All well and good, but what happens *after* yes?

I NEEDED, AS a mom and a journalist, to find out the truth behind the headlines, what was real and what was hype. So I began interviewing girls: engaging in in-depth, hours-long conversations about their attitudes, expectations, and early experiences with the full range of physical intimacy. I recruited daughters of friends of friends (and the friends of those girls, and their friends, too); students of high school teachers I had met. I would ask professors on campuses I visited to send out an e-mail blast, inviting any girls interested in talking to me to get in touch. In the end, I interviewed more than seventy young women between the ages of fifteen and twenty, an age span during which most will become sexually active. (The average American has first intercourse at seventeen; by nineteen, three fourths of teens have had sex.) My focus remained on girls alone because, as a journalist, writing about young women has been a passion, a calling: I've been chronicling their lives for over twenty-five years. Girls, too, continue to live with unique contradictions as they make sexual choices: despite the seismic changes in expectations and opportunity, they're still subject to the same old double standard, the idea that a sexually active girl is a "slut," while a similar boy is a "player." Now, though, girls who abstain from sex, once thought of as the "good girls," are shamed as well, labeled "virgins" (which is not a good thing) or "prudes." As one high school senior said to me, "Usually the opposite of a negative is a positive, but in this case it's two negatives. So what are you supposed to do?"

I don't claim to reflect the experience of all young women. My interview subjects were either in college or college bound—I specifically wanted to talk to those who felt they had all options open to them, the ones who had most benefited from women's economic and political progress. They were also self-selected. That said, I cast my net broadly. The girls I met came from across the country, from large cities and small towns. They were Catholic, mainline Protestant, Evangelical, Jewish, and unaffiliated. Some of their parents were married, some were divorced; some lived in blended families, some in single-parent households. They came from politically conservative as well as liberal backgrounds, though most leaned somewhat toward the latter. The majority was white, but many were Asian American, Latina, African American, Arab American, or mixed race. About 10 percent identified as lesbian or bisexual, though most, particularly those still in high school, had not acted on their attraction to other girls. Two were physically disabled. While most came disproportionately from upper-middle-class families, there was some range of economic background—I interviewed girls from the East Side of Manhattan and the South Side of Chicago; girls whose parents managed hedge funds and those whose parents managed fast-food restaurants. To protect their privacy, I have changed all names and identifying details.

At first, I worried that girls wouldn't discuss such a personal subject with me. I needn't have. Wherever I went, I had more volunteers than I could handle. They were not just eager, they were *hungry* to talk. No adult had ever before inquired about their experience of sexuality: what they did, why they did it, how it felt, what they hoped for, what they regretted, what was fun. Often in interviews, I barely asked a question. The girls would just start talking, and before we knew it, hours had gone by. They told me how they felt about masturbation, about oral sex (both giving and receiving), about orgasm. They talked about toeing that line between virgin and slut. They told me about boys who were aggressive and boys who were caring; boys who abused them and boys who restored their faith in love. They admitted their attraction to other girls and their fears of parental rejection. They talked about the complicated terrain of the hookup culture, in which casual encounters precede (and may or may not lead to) emotional connection; now commonplace on college campuses, it was rapidly drifting down to high school. Fully half the girls had experienced something along a spectrum of coercion to rape. Those stories were agonizing; equally upsetting, only two had previously told another adult what had happened.

Even in consensual encounters, much of what the girls described was painful to hear. Perhaps that seems like nothing new, but that in itself is worth exploring. When so much has changed for girls in the public realm, why hasn't more—*much more*—changed in the private one? Can there be true equality in the classroom and the boardroom if there isn't in the bedroom? Back in 1995 the National Commission on Adolescent Sexual Health declared healthy sexual development a basic human right. Teen intimacy, it said, ought to be "consensual, non-exploitative, honest, pleasurable, and protected against unintended pregnancy and STDs." How is it, over two decades later, that we are so shamefully short of that goal?

Sara McClelland, a professor of psychology at the University of Michigan, writes about sexuality as a matter of "intimate justice," touching on fundamental issues of gender inequality, economic disparity, violence, bodily integrity, physical and mental health, self-efficacy, and power dynamics in our most personal relationships. She asks us to consider: Who has the right to

5

engage in sexual behavior? Who has the right to enjoy it? Who is the primary beneficiary of the experience? Who feels deserving? How does each partner define "good enough?" Those are thorny questions when looking at female sexuality at any age, but particularly when considering girls' early, formative experience. Nonetheless, I was determined to ask them.

A number of the girls I met stayed in touch long after we spoke, e-mailing updates about new relationships or evolving beliefs. "I wanted to let you know that because of our conversation I've changed my major," one wrote. "I'm going to study health with a focus on gender and sexuality." Another, a high school junior, told me our discussion affected the questions she asked while touring college campuses. A third, a high school senior, confessed to her boyfriend that all her "orgasms" had been fake; yet another high-schooler told *her* boyfriend to stop pressuring her to climax; it was ruining sex. The interviews—with the young women themselves and with psychologists, sociologists, pediatricians, educators, journalists, and other experts—changed me, too, forced me to confront my biases, overcome discomfort, clarify my values. That, I believe, has made me a better parent, a better aunt, a better ally to all the young women, and the young men, in my life. I hope, after reading this book, you will feel the same way.

CHAPTER 1

Matilda Oh Is Not an Object— Except When She Wants to Be

Camila Ortiz and Izzy Lang had heard it all before. They were seniors at a large California high school—with a campus of over 3,300 students—so this was their fourth September, their fourth "welcome back" assembly. They sat toward the rear of the auditorium, alternately daydreaming and chatting with friends as administrators droned on about the importance of attendance ("especially for you seniors"); the behaviors that could get you suspended; the warnings about cigarettes, alcohol, and weed. Then the dean of students addressed the girls in the crowd. "He was like, 'Ladies, when you go out you need to dress to respect yourself and respect your family,'" recalled Izzy. Blond and blue-eyed, she had a dimple in one cheek that deepened as she spoke. "'This isn't the place for your short shorts or your tank tops or your crop tops. You need to ask yourself: if your grandmother looks at you, will she be happy with what you're wearing?'"

Camila, whose left nostril was pierced with a subtle crystal stud, jumped in, her index finger wagging. "'You need to cover that up because you need to have respect for yourself.' *You need to respect yourself. You need to respect your family.* That idea was just . . . repeated and repeated. And then he went from that immediately into the slides defining sexual harassment. Like there was a connection. Like maybe if you don't 'respect yourself' by the way you dress you're going to get harassed, and that's your own fault because you wore the tank top."

Growing up in this very school system, Camila had learned the importance of challenging injustice, of being an "upstander." So she began to shout the dean's name. "Mr. Williams! Mr. Williams!" she yelled. He invited her to the front of the auditorium and handed her the mic. "Hi, I'm Camila," she said. "I'm a twelfth-grader and I think what you just said is not okay and is extremely sexist and promoting 'rape culture.' If I want to wear a tank top and shorts because it's hot, I should be able to do that and that has no correlation to how much 'respect' I hold for myself. What you're saying is just continuing this cycle of blaming the victim." The students in the auditorium cheered, and Camila handed back the mic.

"Thank you, Camila. I totally agree," Mr. Williams said as she walked back to her seat. Then he added, "But there's a time and a place for that type of clothing."

This was not the first earful I'd gotten about girls' provocative dress: from parents, from teachers, from administrators, from girls themselves. Parents went to battle over the skimpiness of shorts, the clingy V-necks, the tush-cupping yoga pants that showed "everything." *Why do girls have to dress like that?* moms asked, even as some wore similar outfits themselves. Principals tried to impose decorum, but ended up inciting rebellion. In suburban Chicago, eighth-graders picketed a proposed policy against leggings. High-schoolers in Utah took to the Internet when they discovered digitally raised necklines and sleeves added to female classmates' shirts in their yearbook photos.

Boys run afoul of dress codes when they flout authority: "hippies" defying the establishment, "thugs" in saggy pants. For girls, the issue is sex. Enforcing modesty is considered a way both to protect and to contain young women's sexuality; and they, by association, are charged with controlling young men's. After the assembly, the dean of attendance, who was female, stopped Camila in the hallway. "I totally get that you're trying to empower yourself," she told the girl, "but it's a bit distracting. You have male teachers, and there are male students."

"Maybe you shouldn't be hiring male teachers that are focused on staring at my boobs!" Camila shot back. The dean said they could continue the conversation later. "Later" never came.

That was three months ago, and Camila was still furious. "The truth is, it doesn't matter what I wear," she said. "Four out of five days going to school I will be catcalled, I will be stared at, I will be looked up and down, I will be touched. You just accept it as part of going to school. I can't help my body type, and it's super distracting to *me* to know that every time I get up to sharpen my pencil there's going to be a comment about my butt. That doesn't happen to guys. No guy has ever had to walk down the hall and had girls going, 'Hey, boy, your calves are looking great! Your calves are *hot*.'"

Camila is right. Addressing boys directly is the only way to challenge the assumption by some that girls' bodies exist for them to judge—and even touch—however and whenever they wish. The previous year at the girls' high school, a group of boys created an Instagram account to "expose" the campus THOTs, an acronym for That Ho Over There. (Every generation seems to invent a new *Scarlet Letter* word—*strumpet, hussy, tramp, slut, skank, ho*—with which to demonize girls' sexuality.) They down-loaded pictures from girls' Instagram or Twitter accounts (or snapped one in the hallways), captioning each with the girl's pur-ported sexual history. All the girls singled out were black or La-tina. Camila was one of them. "It was such a violation," she said. "Part of the caption was 'I dare you to go fuck her for a good time.' I had to go to school with that out there." When she lodged a formal complaint, she was placed in a room with four male school security guards who, she said, asked whether she had ac-tually performed the acts attributed to her on the site. Humili-ated, she let the matter drop. The Instagram account eventually petered out; the perpetrators were never caught.

Whether online or IRL ("in real life"), Camila's was hardly an isolated case. Another girl, a high school junior in nearby Marin County, California, who played varsity volleyball, told me how boys from the soccer team gathered in the bleachers to harass her teammates during practice, yelling things like "Nice gooch!"(urban dictionary slang for perineum) when the girls lunged to make a shot. (There are, incidentally, hundreds of close-up, rear-view photos of underage girls in volleyball shorts on the Internet.) A senior in San Francisco described how, within days of arriving at an elite summer journalism program she attended in Chicago, the boys created a "slut draft" (akin to a fantasy football league), ranking their female peers in order of "who they wanted to fuck."

"The girls were pissed off," she told me, "but we couldn't com-plain because of all the implications, right? If you complain and you're on the list, you're a prude. If you complain and you're not on it, you're ugly. If you complain about it being sexist, then you're a humorless feminist bitch and a lesbian."

I heard about a boy who, claiming to have "magic arms,"

would hug random girls in his New York City public school hallway and then announce his assessment of their bra size; about a high school boy who sauntered up to a stranger at a party in Saint Paul, Minnesota, and asked, "Can I touch your boobs?"; about boys at dances everywhere who, especially after a few (or more) drinks, felt free to "grind" against girls from behind, unbidden. Most girls had learned to gracefully disengage from such situations if uninterested. Boys rarely pursued. Several young women, though, said a dance partner had gone further, pushing aside their skirts and sliding a quick finger into their underwear. By college, girls attending a frat party may not make it to the dance floor at all unless they passed what one called the "pretty test" at the front door, where a designated brother "decides whether you are accepted or rejected, beautiful or ugly. He's the reason you better wear a crop top in subzero weather or you'll end up home alone eating microwave popcorn and calling your mom."

I'm going to say this once here, and then—because it is obvious—I will not repeat it in the course of this book: not all boys engage in such behavior, not by a long shot, and many young men are girls' staunchest allies. However, every girl I spoke with, *every single girl*—regardless of her class, ethnicity, or sexual orientation; regardless of what she wore, regardless of her appearance—had been harassed in middle school, high school, college, or, often, all three. Who, then, is truly at risk of being "distracted" at school?

At best, blaming girls' clothing for the thoughts and actions of boys is counterproductive. At worst, it's a short step from there to "she was asking for it." Yet, I also can't help but feel that girls such as Camila, who favors what she called "more so-called provocative" clothing, are missing something. Taking up the right to bare arms (and legs and cleavage and midriffs) as a feminist rallying cry strikes me as suspiciously Orwellian. I recall the simple litmus test for sexism proposed by British feminist Caitlin Moran, one that Camila unconsciously referenced: Are the guys doing it, too? "If they aren't," Moran wrote, "chances are you're dealing with what we strident feminists refer to as 'some total fucking bullshit.' "

So while only girls get catcalled, it's also true that only girls' fashions urge body consciousness at the very youngest ages. Target offers bikinis for infants. The Gap hawks "skinny jeans" for toddlers. Preschoolers worship Disney princesses, characters whose eyes are larger than their waists. No one is trying to convince eleven-year-old boys to wear itty-bitty booty shorts or bare their bellies in the middle of winter. As concerned as I am about the policing of girls' sexuality through clothing, I also worry about the incessant drumbeat of self-objectification: the pressure on young women to reduce their worth to their bodies and to see those bodies as a collection of parts that exist for others' pleasure; to continuously monitor their appearance; to perform rather than to feel sensuality. I recall a conversation I had with Deborah Tolman, a professor at Hunter College and perhaps the foremost expert on teenage girls' sexual desire. In her work, she said, girls had begun responding "to questions about how their bodies feel—questions about sexuality or arousal—by describing how they think they look. I have to remind them that looking good is *not* a feeling." Self-objectification has been associated with depression, reduced cognitive function, lower GPA, distorted body image, body monitoring, eating disorders, risky sexual behavior, and reduced sexual pleasure. In one study of eighth-graders, self-objectification accounted for half the differential in girls' reports of depression and more than two-thirds of the variance in their self-esteem. Another study linked girls' focus on appearance to heightened shame and anxiety about their bodies. A study of twelfth-graders connected self-objectification to more negative attitudes about sexuality, discomfort in talking about sex, and higher rates of sexual regret. Self-objectification has also been correlated with lower political efficacy: the idea that you can have an impact in the public forum, that you can bring about change.

Despite those risks, hypersexualization is ubiquitous, so visible as to be nearly invisible: it is the water in which girls swim, the air they breathe. Whatever else they might be—athletes, artists, scientists, musicians, newscasters, politicians—they learn that they must, as a female, first and foremost project sex appeal. Consider a report released by Princeton University in 2011 exploring the drop over the previous decade in public leadership positions held by female students. Among the reasons these über-elite young women gave for avoiding such roles was that being qualified was not enough. They needed to be "smart, driven, involved in many different activities (as are men), and, in addition, they are supposed to be pretty, sexy, thin, nice, and friendly." Or, as one alumna put it, women had to "do everything, do it well, and look 'hot' while doing it." A 2013 study at Boston College, meanwhile, found that female students were graduating with lower self-esteem than when they entered the school (boys' self-esteem rose). They, too, in part blamed "the pressure to look or dress a certain way." A sophomore in a survey at Duke that reached similar conclusions called the phenomenon "effortless perfection," the "expectation that one would be smart, accomplished, fit, beautiful, and popular, and that all this would happen without visible effort." No wonder they faltered.

"Hot," as journalist Ariel Levy wrote in her book, *Female Chauvinist Pigs*, is different from "beautiful" or "attractive." It is a commercialized, one-dimensional, infinitely replicated, and, frankly, unimaginative vision of sexiness, one that, when applied to women, can be reduced to two words: "fuckable and saleable." Levy says that "hotness" is specifically women's work, and nowhere was that more evident than on the 2015 *Vanity Fair* cover featuring Caitlyn Jenner, née Bruce. To announce her physical transition from male to female, the sixty-five-year-old appeared in a corset (from a store called Trashy Lingerie), breasts overflowing, lips glossed like an ingénue's. That image was often juxtaposed in the press with a picture of her as Bruce, hair lank with sweat, arms raised in triumph after winning Olympic gold. As a man, he used his body; as a woman, she displayed it. Certainly, it's no revelation that girls are held to a punishingly narrow, often surgically or digitally enhanced ideal of "sexy," and then labeled as "sluts" when they pursue it. What has changed is this: whereas earlier generations of media-literate, feminist-identified women saw their objectification as something to protest, today's often see it as a personal choice, something that can be taken on intentionally as an expression rather than an imposition of sexuality. And why wouldn't they, if "hot" has been portrayed as compulsory, a prerequisite to a woman's relevance, strength, and independence?

The girls I met talked about feeling both powerful and powerless while dressed in revealing clothing, using words like *liberating*, *bold*, *boss bitch*, and *desirable*, even as they expressed indignation over the constant public judgment of their bodies. They felt simultaneously that they actively chose a sexualized image—which was nobody's damned business but their own—*and* that they had no choice. "You want to stand out," one college sophomore told me. "You want to attract someone. So it's not just about being hot, but who can be the *hottest*. One of my friends has gotten to the point where she's practically naked at parties." Girls shifted between subject and object day by day, moment by moment, sometimes without intending to, sometimes unsure themselves of which they were. Camila, for instance, had worn a brand-new bustier top to school the previous day. "When I got dressed I was like, 'I feel super comfortable with myself,'" she said. "'I feel really hot and this is going to be a good day.' Then, as soon as I got to school, I felt like"—she snapped her fingers—"automatically I wasn't in control. People are staring at you, looking you up and down, saying things. I started second-guessing myself, thinking, 'I shouldn't have worn this shirt. It's too revealing, it's too tight.' It was dehumanizing." Listening to Camila, I was struck by the assertion that how "hot" she felt would determine the quality of her day; also that, midway through her story, she switched to the second person—as if she, like those around her, suddenly saw herself as an object.

I used to say, when speaking publicly on college campuses or to groups of parents, that one could disentangle sexualization from sexuality by remembering that the first is foisted on girls from the outside, the other cultivated from within. I'm no longer sure it's so simple. It may seem clearly unhealthy when a three-year-old insists on wearing high heels to preschool every day or a five-year-old asks if she's "sexy" or a seven-year-old begs for that padded bikini top from Abercrombie (an item that was pulled from the shelves after parental protest). But what about the sixteen-year-old who washes her boyfriend's car clad in a bikini top and Daisy Dukes? What about that strip aerobics class the college freshman is taking? And what about, you know, that *outfit?* As Sydney, a Bay Area high school senior sporting oversize geek-chic glasses, asked me, "Isn't there a difference between dressing slutty because you *don't* feel good about yourself, and you want validation, and dressing slutty because you *do* feel good about yourself and don't need validation?"

"Could be," I replied. "Explain how you know which is which."

Sydney gazed down at the chipped black polish on her nails and began flipping one of her silver rings from finger to finger and back again. "I can't," she said after a moment. "My whole life is an attempt to figure out what, in the core of myself, I actually like versus what I want to hear from other people, or wanting to look a certain way to get attention. And part of me feels cheated out of my own well-being because of that."

Girls do push back against the constraints of "hot," the contradictory message that it is mandatory yet also the justification for their harassment or assault. A spontaneous movement of "Slutwalks" exploded in 2011, after a Toronto policeman suggested that college women who wanted to avoid sexual assault shouldn't dress so provocatively. Infuriated, young women across the globe, many in fishnets and garters, hit the streets bearing signs reading such things as "My Dress Is Not a Yes!" and "My Ass Is Not an Excuse for Assault!" At the other end of the spectrum, Generation Y made news both by growing out their armpit hair and rejecting the torture device commonly known as thong underwear (some in favor of "granny pants" with "Feminist" stamped across the rump), proving they could be sexy without pandering to "hot." On a more personal level, one of the young women I met, an art student, told me that, tired of the "costume" that girls were expected to don at college parties, she was opting for a different one, showing up dressed as a sparkly unicorn. "I feel liberated," she told me. "It's still kind of body-conscious, and there is a lot of makeup involved, but I'm also fully covered. And I'm one-of-a-kind."

Hot or Not: Social Media and the New "Body Product"

Girls did not always organize their thinking about themselves around the physical. Before World War I, self-improvement meant being *less* self-involved, *less* vain: helping others, focusing on schoolwork, becoming better read, and cultivating empathy. Author Joan Jacobs Brumberg highlighted this change in her book *The Body Project* by comparing the New Year's resolutions of girls at the end of the nineteenth and twentieth centuries:

"Resolved," wrote a girl in 1892, "to think before speaking. To work seriously. To be self-restrained in conversations and actions. Not to let my thoughts wander. To be dignified. Interest myself more in others."

And one hundred years later:

"I will try to make myself better in any way I possibly can. . . . I will lose weight, get new lenses, already got new haircut, good makeup, new clothes and accessories."

Brumberg's book was published in the late 1990s, a good decade before social media took off. With the advent of MySpace, then Facebook, then Twitter, Instagram, Snapchat, Tumblr, Tinder, YikYak, and—mark my words—some social media–linked microchip that they'll soon implant in all our heads, the body has become even more entrenched as the ultimate expression of the female self, evolving from "project" to consciously marketed "product." There are myriad ways social media can be fun, creative, connective, political. They can be a lifeline for kids who feel different from their peers, particularly LGBTQ teens, providing them with crucial support and community. They have also reinforced the relentless externalization of girls' sense of self. There is evidence that the more concerned a girl is about her appearance, weight, and body image, the more likely she is to consult the magic mirror of her social media profile, and vice versa: the more she checks her profile, the more concerned she becomes about appearance, weight, and body image. Comments on girls' pages, too, tend to focus disproportionately on looks, and even more than in the real world, that becomes a measure of friendship, self-image, and self-worth.

In a windowless basement office on a private midwestern college campus, Sarah, a first-semester sophomore, stood in front of me with the toes of one foot pointed forward, one knee slightly bent, to demonstrate the "leg bevel"—a pose pioneered by showgirls but which is now standard in girls' social media photos. "It slims your body more than if you stand normally," she explained. Sarah grew up in Atlanta, where she attended a small Christian high school. She had dyed blond hair that hung to her shoulders, blue eyes, and carefully applied makeup—foundation, eye shadow, lipstick. "People will"—she stopped and laughed self-consciously—"this is so stupid, but people will learn the ways to pose in pictures so they'll look good on Facebook or Instagram. I mean, I do it. A hand on your hip—that makes you look thinner, too. Or, whichever side you part your hair on, the other side would be your 'better' side, so I try to face this way in photos." She turned her right cheek toward me and continued. "I edit little blemishes out and fix the lighting. And if you watch things like *America's Next Top Model*, you learn to 'find your light.' Things like that."

Teens have always been acutely aware of how they are seen by their peers. Social media amps up that self-consciousness: rather than experimenting among a small group of people they actually know, they now lay out their thoughts, photos, tastes, and activities (as well as their lapses in judgment) for immediate approval or censure by their 947 BFFs, many of whom are relative strangers. The result, according to Adriana Manago, a researcher at the Children's Digital Media Center in Los Angeles who studies college students' behavior on social media, is that young people have begun talking about the self as a brand rather than something to be developed from within. Their "friends" become an audience to be sought after and maintained. Ninety-two percent of teens go online daily, including 24 percent who are online "almost constantly." Nearly three-quarters use two or more social networking sites. Also, especially on photo-sharing sites such as Instagram, girls are more active than boys, who are more likely to be gamers. "You use your experience to create an image," Matilda Oh, a high school senior in San Francisco told me, "with the ultimate goal being to show that you're desirable and attractive and wanted and liked." Every young woman, she said, knows that she will "get ten times as many 'likes' by posting a picture of yourself in a bikini than you would if you were wearing a snow jacket." Yet, just as in the real world, girls must be careful to come off as "hot" yet not "slutty," sexually confident but not "thirsty." In one study of 1,500 Facebook profiles, college-age women judged other girls' profiles far more harshly than they did boys', criticizing those who had "too many" friends, shared "too much" information, showed "too much" skin in photos, name checked their boyfriends "too often," posted "too many" status updates. This despite the fact that 1,499 of the profiles aspired to the same "ideal": a girl who, through status updates, glamour shots, and skin-bearing selfies, depicted herself as "fun" and "carefree"; who had lots of attractive friends, went to lots of parties, and was interested mostly in romance, pop culture, and shopping. You could easily get trashed, then, for the very thing you needed to do to court approval.

It doesn't take much to become a target. "You can totally

get stigmatized," agreed Sarah. "I knew a girl who only Insta-grammed selfies. Every single picture was a selfie. And people talked about it. It made her seem like she either had no friends or was too into herself. There are so many ways to be judged. And of course you're afraid that the judgments you pass against others will be passed against you. It's not something you ever talk about, though. You just try to listen to what people say and kind of learn those unwritten rules. Like, don't change your profile picture too much. Don't post statuses about everything you're doing. Don't have too many pictures of yourself."

In 2013 *selfie* was named the "international word of the year" by Oxford Dictionaries. Anyone with a Facebook or Instagram account probably has posted a few, but no one matches the self-chronicling output of adolescent girls (interestingly, after age forty, men become the more dominant selfie posters—perhaps in midlife, women unconsciously render themselves invisible). The portraits can be a giddy assertion of pride for young women, a claim staked for the right to take up public space. "If you write off the endless stream of posts as image-conscious narcissism," Rachel Simmons, author of *Odd Girl Out*, has written, "you'll miss the chance to watch girls practice promoting themselves—a skill that boys are otherwise given more permission to develop, and which serves them later on when they negotiate for raises and promotions."

Personally, I love flipping through posts by the girls I know (my nieces, my friends' kids, the girls I interview), seeing them in front of national monuments or on graduation day or clown-ing around with friends. That doesn't, however, allay my con-cern that selfies can impose another tyranny on girls, another imperative to dish up their bodies for inspection by others and themselves, another way in which their value is reduced to the superficial, flattened, measured by visibility. As one girl said to me, "It's like cell phones, Facebook—all of it comes back to the issue of: Am I pretty? How many friends do I have? How do my profile pictures look? Let me stalk *myself*."

The girls I met, again, were not passive; they were not victims of social media. They were acutely literate, often avidly feminist. They actively engaged in contemporary culture even as they struggled with the meaning and impact of that engagement. Nearly two-thirds of teen girls in one large-scale survey did feel that selfies boosted their confidence. So there's that. But about half also said that photos posted of them by *others* (presumably less mindful of their best angles) have the potential to make them feel bad. Body dissatisfaction seems less driven by the actual time young women spent on social media than by how much of that time they spend sharing and viewing photos; the more they look at others' pictures, whether close friends or distant peers, the more unhappy they become about their own appearance. Little wonder, then, that there has been a proliferation of "selfie surgery apps," which allow a user to shrink her nose, whiten her teeth, broaden her smile. Actual plastic surgery among those under thirty is on the rise, too. In 2011 there was a 71 percent increase in the number of high school girls obtaining chin implants specifically because they wanted to look better in prom selfies. One of every three members surveyed by the American Academy of Facial Plastic and Reconstructive Surgery in 2013 said that their patients sought their services to look better in selfies.

Posting pictures of yourself—even lots and lots of pictures of yourself—while eating cereal or shopping for a prom dress or hanging with your besties is one thing. What really worries parents is the selfie's evil cousin: the sext. *Do not*, we tell our daughters, absolutely *do not* send anyone sexually explicit messages or, God forbid, a nude or seminude photo. The Internet is forever, we say. Snapchat doesn't prevent screenshots that can be redistributed in an instant and used as weapons (witness the rise of "revenge porn": explicit images posted online without the victim's consent, often following a breakup). In truth, it's hard to know exactly how common "sexting" is among teens. In surveys, between 15 and 48 percent (depending on the age of the children asked and how "sexting" is defined) say they have sent or received an explicit text or photo. What is clear, though, is that the practice is not gender neutral. While equal numbers of boys and girls may sext voluntarily, girls are twice as likely to be among those who were pressured, coerced, blackmailed, or threatened into it—fully half of teen sexting in one large-scale survey fell into those categories. That's particularly disturbing, since coercion into sexting appears to cause more long-term anxiety, depression, and trauma than coercion into real-life sex. Among the girls I met, the badgering to send nude photos could be incessant, beginning in middle school. One girl described how, in eighth grade, a male classmate threatened (in a text) to commit suicide if she didn't send him a picture of her breasts. She told her parents, while a friend of hers he also targeted complied. Sometimes the pressure was mixed with girls' own desire to please, to provoke, or to be affirmed as hot. They sexted photos to boyfriends to prove their trust, or to boys whose interest they hoped to attract. (Boys did this, too, but girls typically considered it aggressive and "gross.") One girl told me that there had been an "epidemic" of classmates at her private Jewish middle school who flashed their breasts at boys while video chatting. The boys began taking screenshots and posting them online.

"Did the girls want that to happen?" I asked.

"No," she said. "But it did." By high school, the girls had "grown out of it," but the boys had not. "I would video chat with boys, and they'd be like, 'Come on! Flash me! Flash me!' I wouldn't do it, but they'd be very persistent. They'd say, 'Just do it. I promise I won't take a picture.' And if you really like the guy, you think maybe he'll like you back. . . . There were boys who had whole folders of pictures. Like trophies."

Some girls considered sexting and sexy video chatting a way to experiment with sex safely (at least as they saw it). "I would do really graphic sexting over IM in middle and high school," a freshman at a mid-Atlantic college told me, "or do stripteases on Skype. I wasn't ready to lose my virginity, but I loved being the bad girl." She didn't worry that her recipients might share her performances; she believed she could use her body to intimidate as well as entice. "I'm six feet tall," she said. "I'm not this dainty little thing. I was like, if you pass this around you will not have balls anymore. I will *hurt* you. So I felt in control."

Are selfies empowering or oppressive? Is sexting harmful or harmless? Is that skirt an assertion of sexuality or an exploitation of it? Try this: looking up at the ceiling, raise your hand over your head and trace a clockwise circle with your index finger. Continue to trace the circle while slowly lowering your arm so that your finger is at eye level. Now, still tracing, lower your arm further until it's at your waist. Look down at the circle. Which direction is it spinning? Although it would seem impossible, the circle moves clockwise and counterclockwise at the same time. Management consultants use that "both/and" concept to break down rigid "either/or" thinking. Deborah Tolman has suggested that it's equally useful when considering young women's complicated relationships to their bodies, their sexuality, and sexualization. That's the challenge to both parents and girls themselves: whether you're discussing dress codes, social media, or the influence of pop culture, there is rarely a clear-cut truth.

Parts Is Parts

2014 was "all about that bass," the lyrics to Meghan Trainor's wildly popular confection themselves riddled with both/and contradiction. The song ostensibly celebrated the body positive, rejecting the "stick-figure silicone Barbie doll" ideal. Yet it contained a Trojan horse: not only did Trainor take a gratuitous swipe at "skinny bitches" (followed by a coy "No, I'm just playing"), but she also reassured young women that "boys, they like a little more booty to hold at night." So, sure it's fine to be curvier—as long as guys still think you're hot.

Trainor was kind of late to the party, though: the "bass" was already in ascendance, metamorphosing from a Sir Mix-A-Lot novelty song to a JLo trademark, to a national obsession. On the cover of her single "Anaconda," Nicki Minaj squatted, back to the camera, her knees splayed, revealing a prodigious (and, rumor has it, surgically enhanced) posterior. The art for Lady Gaga's single "Do What You Want" featured a be-thonged upthrust rear. (The chorus of the song itself, a duet with alleged child rapist R. Kelly, is "Do what you want, what you want with my body.") During her On the Run tour, Beyoncé appeared in a Givenchy-designed bodysuit with cutouts that showed off her naked buns. The cover of the 2014 *Sports Illustrated Swimsuit Issue* portrayed yet another rear view: three topless supermodels gazing playfully over their shoulders as they offer their near-bare bum cheeks for readers' inspection. Later that year, Lopez rereleased her trend-launching hit "Booty" with a new, far more explicit video, featuring "Pu$$y" rapper Iggy Azalea. And Kim Kardashian notoriously tried to "break the Internet" with a *Paper* magazine cover shot of her bounteous (and, again, possibly augmented) derrière, slick with baby oil.

And there's more! Jen Selter, a fitness model dubbed the Belfie Queen—that's "butt selfie"—has over 7 million Instagram oglers and earns as much as sixty thousand dollars for sponsored posts. For the more ordinary mortals, an eighty-dollar gizmo called a "belfie stick," designed to help capture that perfect rear angle, sold out immediately online and, at this writing, had a months-long waiting list. Between 2012 and 2013 the number of "Brazilian butt lifts" performed in the United States, in which fat is transferred from another part of the body to the rear, jumped 16 percent. For those short the ten thousand dollars required for that procedure, sales of twenty-two-dollar Booty Pop panties—think padded bra for the bum—were up in November 2014 nearly 50 percent from the same period a year earlier; the company subsequently introduced a new, larger product, with 25 percent more foam.

Maybe it was just the butt's turn: After all, how many more hours could women while away obsessing over their stomachs, breasts, hips, upper arms, necks, and faces? How many more cosmetic procedures could they undergo? Something had to fill the breach. Truly, you'd think that after buying into the horror of the "thigh gap," women would resist being defined by yet another body part, particularly this one. As Amy Schumer rapped in "Milk, Milk, Lemonade," her brilliant send-up of the booty craze, we're "talkin' 'bout my fudge machine." The girls I met, though, didn't see it that way. Matilda Oh suggested I was hypocritical for dismissing Nicki Minaj as self-objectifying in "Anaconda" but hailing Lena Dunham as subversive for playing Ping-Pong topless on *Girls*. But Dunham wasn't trying to be hot. Quite the opposite: she is dough-bellied and soft-chinned, with natural, lopsided breasts. Her "bass" is perhaps a little *profundo*. In other words, she looks like an average American woman. She uses her body to shatter taboos against showing the imperfectly ordinary, to challenge our pneumatic, implant-propelled expectations. "Nicki Minaj is challenging, too," Matilda countered. Minaj cast off shame, rejected the male-generated shackles of "respectable" female sexual behavior, refused to see the behind—especially when it's large, especially when it's attached to a woman of color—as "dirty." "People always gripe about Nicki Minaj's butt," Matilda said, "but I think it's kind of 'damned if you do, damned if you don't.' If you emphasize it, you could potentially normalize black bodies in the mainstream, but you'll also be accused of 'objectifying' yourself. If you don't do it, though, you are arguably participating in a culture of body shame. So how is a woman of color supposed to 'take control of her sexuality' or be 'body-positive' without it being construed as internalized fetishization?"

Is Minaj's butt transgressive? What about Gaga's? What about those *Sports Illustrated* swimsuit models'? How can one tell which of these images is defiant and which is complicit; which liberates and which limits; which undermines standards of beauty and which creates new ones? Can they do both simultaneously? "I love Beyoncé," a freshman at a West Coast college told me. "She's one of my idols. She's, like, a queen. But I wonder, if she wasn't beautiful, if people didn't think she was so sexy, would she be able to make the feminist points she makes?"

Feminist scholar bell hooks, who kicked the Beyhive in 2014 when she called Beyoncé a "terrorist, especially in terms of her impact on young girls," has suggested that the fascination with the butt is nothing more than the latest way to reduce a woman to a body part: the latest PG-13 stand-in for "the pussy." The obsession is no different, no more subversive, and no more "empowering" to women than the fetishization of the breast or the wet, open mouth. As with those pop culture memes, she said, it raises the basic question: "Who possesses and who has rights to the female body?"

Young fans such as Matilda argue that the stars themselves do. Female artists, they insist, are taking control (or at least are being marketed as taking control) of a hypersexualized industry that too often exploits women. Yes, these women may be products, but they are also *producers*. The decision to twerk onstage, or twirl on a poll, or dance in one's drawers around a fully clothed man, or to pose nude on the cover of a magazine is now a woman's alone: rather than capitulating, they are actually reclaiming their sexuality. Yet those performers still work within a system that, for the most part, demands women look and present their bodies in a particular way in order to be heard, in order to be seen, in order to work. Successfully manipulating that system to their advantage by, say, nominally reimagining the same old strip club clichés may get them rich, it may get them famous, but it shouldn't be confused with creating actual change. Artists such as Gaga or Rihanna or Beyoncé or Miley or Nicki or Iggy or Kesha or Katy or Selena may not be puppets, but they aren't necessarily sheroes, either. They're shrewd strategists, spinning commodified sexuality as a choice, one that may be profitable but is no less constraining, ultimately, either to female artists or to regular girls. So the question is not whether pop divas are expressing or exploiting their sexuality so much as why the choices for women remain so narrow, why the fastest route to the top as a woman in a sexist entertainment world (just as for ordinary girls on social media) is to package your sexuality, preferably in the most extreme, attention-getting way possible.

The Twerk Seen 'Round the World

Miley Cyrus's face floated against the back wall of Oakland's Oracle Arena, a cross between a humongous selfie and the disembodied head from *The Wizard of Oz*. An eye winked, the lips pursed and stretched. A pink tongue unfurled, and suddenly the real Miley, dressed in a red, spangly two-piece leotard with bird feather shoulders, stepped out, arms aloft, and slid onto the stage. As she launched into the opening lines of her song "SMS (Bangerz)," tens of thousands of girls (and a few boys) screamed and held up flashing iPhones, the latter-day version of waving cigarette lighters. It was February 2014, about six months since Miley buried her Disney image forever with what has been called "the twerk seen 'round the world."

For those who may have recently migrated from Pluto, Miley sparked international outrage with her performance at the 2013 MTV Video Music Awards, first mimicking anilingus on a black female backup dancer (who inexplicably had a giant stuffed teddy bear strapped to her back) and then stripping down to plastic, skin-colored skivvies and vibrating her butt, or "twerking," against Robin Thicke's crotch as the two performed his controversial hit, "Blurred Lines." She also employed a foam finger, the kind fans typically brandish at sporting events, in ways that, once witnessed, could never be unseen. Throughout, there was that freakish wagging tongue, which now rivals KISS's Gene Simmons's in notoriety. The performance sparked predictable handwringing by both conservative pundits and feminists. (No less than Sinead O'Connor, who once shredded a picture of the Pope on live TV while singing the word *evil*, urged Miley "not to let the music industry make a prostitute of you.") Then came a backlash, led largely by young women, accusing both groups of "slut-shaming" Miley for "expressing her sexuality." Miley was also attacked as racist for appropriating aspects of black "ratchet" culture to boost her bad-girl cred and using her voluptuous backup dancers' bodies as props. None of it mattered. By the next morning Miley's singles had secured the top two spots on iTunes. The *Bangerz* album, released about six weeks later, debuted at number one on the *Billboard* charts.

This was not my first Miley concert. Five years earlier I'd attended her Wonder World tour, also at the Oracle Arena, where she shocked a crowd of tweenie-bopper *Hannah Montana* fans by grinding on the boys in the band while clad in leather short shorts and a cleavage-baring vest. This time around it seemed those little girls (or their mothers) had gotten the memo: they were nowhere in sight. Or maybe they were here; they were simply older now, as was Miley. Before the show, the halls of the arena were jammed with young women in their late teens and early twenties sporting Miley's pigtail bun hairstyle from the VMAs. Some wore crop tops emblazoned with the word *twerk* in six-inch-high capital letters. A number of them carried foam fingers. A few had found knock-offs of the furry teddy-bear-faced lingerie teddy that Miley had worn before stripping at the awards, with its winking eye and chops-licking tongue. ("Miley Cyrus costume" had been the second most popular search on Google that Halloween, with those teddies selling for around ninety dollars a pop.) One girl waltzed by clad in a flesh-toned bra and panties, and while that might not in itself have raised eyebrows, the two middle-aged adults trailing her with cameras, paparazzi-style (presumably her parents), turned more than a few heads. There was lots of midriff on parade, a lot of leg, a lot of stilettos. The smell of weed permeated the air.

I planted myself near a concession stand, where no fewer than thirty girls over the course of about fifteen minutes asked me to snap their picture beside a life-size poster of Cyrus displaying her famous tongue. A few made "duck lips" or "faux surprise" face—*I'm fun! I'm ironic!*—but most imitated their idol. I asked one girl, a nineteen-year-old named Emilia, to explain the appeal of the pose. "I guess it's to say, 'I don't care,'" she said.

"You don't care about what?"

She shrugged. "I just *don't care!*"

A twenty-one-year-old women's studies major from San Francisco State University stood nearby dressed in a black-and-white striped romper, her hair wound into pigtail buns, a slash of red lipstick on her mouth. "I like Miley because she is just herself," she explained. "I loved *Hannah Montana*. I've seen every episode. But I'm grown up now, and so is Miley. She needed to break free and show that she wasn't the Disney star anymore." The girl looked around the hallway. "And she did."

"She is the epitome of perfection," enthused her friend. "And she's *not* going to fit into any cultural ideal. Everyone tells you who you're supposed to be as a girl, but Miley? She is just who she is."

The show itself was a kaleidoscope of quasi-psychedelic images. A caricatured animated Miley (conceived by *Ren and Stimpy* creator John Kricfalusi), bug-eyed and buck-toothed, with huge, flopping butt cheeks, cavorted on-screen as the real-life version performed with those plush stuffed dancing bears, pinching and palpating more backup dancers. A giant bed disgorged dancers of both sexes, who joined Miley in a mock orgy. She simulated intercourse with a "little person," pantomimed fellatio on a dancer dressed as Abraham Lincoln ("party in the USA!"). She urged her audience to make out with each other, drawling, "The more tongue, the better. The *dirtier*, the better." The "nastiest" couples, she said, would be projected onto Jumbotrons flanking the stage. ("Girl on girl is *always* appreciated," she said with a smirk.)

The show was unquestionably graphic but not especially erotic. The images and actions were too random, too devoid of larger meaning or purpose. They were just so much flotsam and jetsam seemingly thrown out to stimulate reaction—*any* reaction. *Look, a thirty-foot cat! Miley in a cannabis body suit! Miley getting off on the hood of a car! Miley astride a giant, airborne hot dog!* Cyrus, her pixie-cut hair dyed platinum, was thinner than she'd been five years before, with no curve to her hips or breasts. She looked surprisingly androgynous, an R-rated Cathy Rigby, a tripped-out Peter Pan. Watching her, I recalled Ariel Levy's observation that Paris Hilton was the perfect celebrity for a time when interest in the *appearance* of sexiness had surpassed interest in the existence of sexual pleasure. In Hilton's famed sex tapes, she looks excited only when posing for the camera; during the actual sex, she seems bored, even taking a phone call in the midst of intercourse. Today's "raunch culture," Levy wrote, is not liberating or progressive, not about "opening our minds to the possibilities and mysteries of sexuality." There's a disconnect between its representation of "hotness" and sex itself. Even Hilton, Levy pointed out, has said, "My boyfriends always tell me I'm not sexual. Sexy, but not sexual."

Maybe Miley provides a release for her fans, an escape from respectability, a vision, however compromised, of a girl who doesn't dither over whether anyone (parents, other entertainers, the media) thinks she is "too slutty." The crotch slapping, the butt shaking, the crude talk, the simulated sex acts—all gave the illusion of sexual freedom, the illusion of rebellion, the illusion of defiance, the illusion that she "doesn't care." But of course Miley *does* care. As someone trying to maintain her status as a celebrity, as a chart topper, she cares very, very much. I keep coming back to her because I find her not unique, but the opposite: she's a human Rorschach, a lint trap of images and ideas about mainstream, middle-class girlhood. When she was fifteen, that meant wearing a "purity ring" and vowing to be a virgin until marriage; at twenty-three, it meant performing mechanistic, mock sex acts on a dwarf while dressed in a racy leotard emblazoned with pictures of money—and calling that liberation. Perpetually striving to mix the perfect cocktail in her cultural blender, she both reflects and rejects what a young woman needs to do to maintain celebrity, to snag attention, to be noticed, to be "liked"—all without seeming to try. And isn't that what every girl is struggling to do, writ very, very large?

In the middle of the show, Miley took a break from singing to address the audience. "How the fuck are you?" she bellowed. Then she turned around, lifted her iPhone high over her head, stuck out her tongue, snapped a selfie with the crowd as backdrop, and posted it immediately to Instagram. She was, it seemed, just like them.

Pop Goes the Porn

"I'm very sensitive about porn," said Alyson Lee, tugging nervously at her dark, purple-streaked hair. Alyson was nineteen, a sophomore at a mid-Atlantic college. She'd grown up in what she called a "culturally conservative" Chinese family in a Los Angeles suburb made up almost entirely of immigrant parents and first-generation kids like her. She studied how Americans are supposed to act and feel, especially about sex and romance, by watching *Grey's Anatomy.* "So now," she said, "I have the very typical, liberal college woman point of view."

One that includes ambivalence about porn. Alyson has had two serious boyfriends—one in her senior year of high school, one as a freshman in college—and both told her the same thing: "Of course I watch porn: every teenage boy does."

"I'm not one of those people who think that porn is wrong and morally terrible and disgusting," Alyson explained. "But it makes me feel super insecure. Like, am I not good enough? I'm definitely not as hot as a porn star. And I'm not going to do the things porn stars do. Both guys were really reassuring that it wasn't about me, and I knew logically there was no connection between them watching porn and some flaw of mine. But it stayed in my head."

If, as bell hooks suggested, pop culture portrayals of women beg the question "Who has access to the female body?" the answer may ultimately be found in the ever-broadening influence of porn. That is, after all, the source of the arched backs, the wet open mouths, the ever-expanding breasts and butts, the stripper poles, the twerking, and the simulated sex acts. That is the source of women's sexuality as a performance for men.

The Internet has made porn more prevalent and accessible than at any time in history, especially to teens. As with pop culture, that has spurred an escalation in explicitness, a need to push the boundaries to maintain an easily distracted audience. Mirroring (and raising further questions about) the mainstream culture's "booty" trend, in one large-scale study of sexual behavior and aggression in best-selling porn, anal sex was depicted in over half of the videos surveyed, always as easy, clean, and pleasurable to women; 41 percent of videos also included "ass to mouth," in which, immediately after removing his penis from a woman's anus, he places it in her mouth. Scenes of "bukake" sex (multiple men ejaculating on one woman's face), "facial abuse" (oral sex aimed at making a woman vomit), triple penetration, and penetration by multiple penises in a single orifice are also on the rise. I'm going to go out on a limb here and say that in real life those practices wouldn't feel good to most women. Watching natural-looking people engaging in sex that is consensual, mutually pleasurable, and realistic may not be harmful— heck, it might be a good idea—but the occasional feminist porn site aside, that is not what the $97 billion global porn industry is shilling. Its producers have only one goal: to get men off hard and fast for profit. The most efficient way to do so appears to be by eroticizing the degradation of women. In the study of behaviors in popular porn, nearly 90 percent of 304 random scenes contained physical aggression toward women, while close to half contained verbal humiliation. The victims nearly always responded neutrally or with pleasure. More insidiously, women would sometimes initially resist abuse, begging their partners to stop; when that didn't happen, they acquiesced and began to enjoy the activity, regardless of how painful or debasing it was. The reality is, as one eighteen-year-old pursuing a porn career told documentary filmmakers Jill Bauer and Ronna Gradus, "I'm supposed to be having sex with guys I would never have sex with, and saying things I would never say. There is nothing sexually arousing. You're just processed meat."

Media has been called a "super peer," dictating all manner of behavioral "scripts" to young people, including those for sexual encounters: expectations, desires, norms. In one era, they learn that you don't kiss until the third date; in another, they learn that sex precedes an exclusive relationship. Bryant Paul, a professor of telecommunications at Indiana University Bloomington who studies "scripting theory," explained, "I'll ask students, 'Think about how you learned what to do at your first college party. You'd never been to one, but you knew you were supposed to gather around the keg. You knew that couples would go off to someone's room.' And they'll say, 'Yeah, from *American Pie* and all those movies.' So where are they learning their sexual socialization, especially in terms of more explicit behaviors? You'd be foolish not to think they're getting ideas from porn. Young people are not *tabulae rasae*. They have a sense of right and wrong. But if they're repeatedly exposed to certain themes, they are more likely to pick them up, to internalize them and have them become part of their sexual scripts. So when you see consistent depictions of women with multiple partners and women being used as sex objects for males, and there's no counterweight argument going on there . . ." He trailed off, leaving the obvious conclusion unspoken.

Over 40 percent of children ages ten to seventeen have been exposed to porn online, many accidentally. By college, according to a survey of more than eight hundred students titled "Generation XXX," 90 percent of men and a third of women had viewed porn during the preceding year. On one hand, the girls I met knew that porn was about as realistic as pro wrestling, but that didn't stop them from consulting it as a guide. Honestly? It pains me to hear that the scatological fetish video *Two Girls, One Cup* was, for some, their first exposure to sex. Even if what they watch is utterly vanilla, they're still learning that women's sexuality exists for the benefit of men. So it worried me to hear an eleventh-grader confide, "I watch porn because I'm a virgin and I want to figure out how sex works"; or when another high-schooler explained that she "watched it to learn how to give head"; or when a freshman in college told me, "There are some advantages: before watching porn, I didn't know girls could squirt."

There is some indication that porn has a liberalizing effect: heterosexual male users, for instance, are more likely than peers to approve of same-sex marriage. On the other hand, they're also less likely to support affirmative action for women. Among teenage boys, regular porn use has been correlated with seeing sex as purely physical and regarding girls as "play things." Porn users are also more likely than their peers to measure their masculinity, social status, and self-worth by their ability to score with "hot" women (which may explain that disproportionate pressure girls report to text boys naked photos of themselves as well as the plots of most Seth Rogen films). Male *and* female college students who report recent porn use have been repeatedly found to be more likely than others to believe "rape myths": that only strangers commit sexual assault or that the victim "asked for it" by drinking too much or wearing "slutty" clothing or by going to a club alone. Perhaps because it depicts aggression as sexy, porn also seems to desensitize women to potential violence: female porn users are less likely than others to intervene when seeing another woman being threatened or assaulted and are slower to recognize when they're in danger themselves.

Boys (both in high school and college), not surprisingly, use porn more regularly than girls. Slightly less than half of male college students use it weekly; only 3 percent of females do. Given that frequent consumers of porn are more likely to consider its depictions of sex realistic, this can skew expectations in the bedroom. "I do think porn changes how guys view sex," mused Alyson Lee. "Especially with my first boyfriend. He had no experience. He thought it would happen like in porn, that I'd be ready a lot faster and he could just, you know, *pound*."

"They think they're supposed to do this hammer-in-and-out thing and that's what girls like," agreed a sophomore at a California college. "They don't realize, 'Dude, that does *not feel good.*' It's all they know. It's what they see. If you're just hooking up with someone, like a one-time thing or whatever, you just *pretend* it feels good."

In her prescient book *Pornified*, Pamela Paul found that women had begun feeling competitive with porn stars, worried that unless they put on their own show to maintain a partner's interest, they would lose him to the Internet. They believed that the unnatural thinness, inflated breasts, and overfilled lips of those surgical cyborgs were distorting men's standards of beauty, eroding women's own body image, increasing their self-consciousness. "Porn has terrible effects on what young women are supposed to look like, particularly during sex," said Leslie Bell, a psychotherapist and author of *Hard to Get: Twenty-Something Women and the Paradox of Sexual Freedom.* "There's this idea that someone is going to be evaluating your appearance not only outside of the bedroom, which was true before, but also during sex, that your body has to look a certain way then. It seems very pressured and shame-inducing, because bodies don't look like that naturally." You'd need self-esteem of steel to remain immune.

The girls I met sometimes disconnected from their bodies during sex, watching and evaluating their encounters like spectators. "I'll be hooking up with some guy who's really hot," confided a high school senior in Northern California, "and we'll be snuggling and grinding and touching and it's cool. Then things get heavier and all of a sudden my mind shifts and I'm not a real person: it's like, *This is me performing. This is me acting.* It's like,

How well am I doing? Like, *This is a hard position, but don't shake.*
And I'm thinking, *What would 'she' do? 'She' would go down on him.'*
And I don't even know who it is I'm playing, who that 'she' actu-
ally is. It's some fantasy girl, I guess, maybe the girl from porn."

JON MARTELLO IS a simple guy, a New Jersey native who
cares about "my body, my pad, my ride, my family, my church,
my guys, my girls, my porn," not necessarily in that order. The
protagonist of the film *Don Jon*, played by Joseph Gordon-Levitt
(who also wrote and directed), Jon Martello got his womanizing
nickname by "pulling" a different girl every weekend. No single
partner, though, can compare to the bounty he finds online. "All
the bullshit fades away," he says in a voice-over, "and the only
thing in the world is those tits . . . dat ass . . . the blow job . . . the
cowboy, the doggie, the money shot and that's it, I don't gotta say
anything, I don't gotta do anything. I just fucking lose myself."

During one of Jon's after-Mass Sunday dinners at his parents'
house, where the TV inevitably blares in the background, an
ad for Carl's Jr. comes on. The camera lingers as Nina Agdal,
a *Sports Illustrated* swimsuit model, rubs oil onto her glistening,
bikini-clad body. She arches like a cat on hands and knees on a
beach, hair flowing, and then sits, legs splayed, and takes a big,
juicy bite of a codfish sandwich. Don's mother looks away, fid-
dles with an earring. His sister, whose back is to the TV, doesn't
even glance up from her cell phone. Don and his father, wearing identical white tank tops, stare at the screen wearing identical slack-jawed expressions. In an interview about that scene,
Gordon-Levitt commented, "What I'm saying is whether it's rated
X or 'approved by the FCC for a general viewing audience,' the
message is the same."

He's right. You don't need to log in to PornHub to absorb its scripts; they're embedded in the mainstream. And the impact of that garden-variety, "pornified" media on young people—from *Maxim* magazine to Dolce & Gabbana couture ads to *Gossip Girl* to multiplayer online games to infinite music videos—is indistinguishable from actual porn. The average teenager is exposed to nearly fourteen thousand references to sex each year on television; 70 percent of prime-time TV now contains sexual content. College men who play violent, sexualized video games are more likely than their peers to see women as sex objects as well as to be more accepting of rape myths, more tolerant of sexual harassment, and to consider women less competent. College women who, in experiments, played the virtual game *Second Life* using a sexualized avatar were more likely to self-objectify *off*line than those who played with a nonsexualized avatar and, again, were more likely to hold false beliefs about rape and rape victims. (Seeing yourself as an object apparently leads you to view other women that way as well.) Meanwhile, in a study of middle and high school girls, those who were shown sexualized pictures of female athletes subsequently scored higher on measures of self-objectification than those who saw the same athletes engaged in, you know, actual athletics. Young women who consume more objectifying media also report more willingness to engage in sexualized behavior, such as taking a pole dancing class or entering a wet T-shirt contest, and to find those activities empowering. They're more likely to justify sexualization, and less likely to protest against it. In other words, as Rachel Calogero, a psychologist at the University of Kent in England, has written, "Objects don't object."

The sex in TV and movies can be simultaneously explicit and evasive. Sex, particularly noncommitted sex, is typically presented

as fun and advisable; rarely is it awkward or silly or challenging or messy or actively negotiated or preceded by discussion of
contraception and disease protection. There's always plenty of
room in the backseat of those limousines, and nary a pothole in
the road. Of course there are exceptions: *Glee* in its early seasons
deftly portrayed issues surrounding teen pregnancy, sex and dis-
ability, homosexuality, bisexuality, first intercourse, fat and slut
shaming, and the nature of love. *Orange Is the New Black*, which was
beloved by many of the girls I met, brought unprecedented gender and sexual diversity to TV. The sex in Lena Dunham's work
is radically raw. One of the most realistic (if depressing) scenes
ever filmed may be found in her 2010 release, *Tiny Furniture*. In
it Aura, a newly minted college graduate played by Dunham, finally gets together with the object of her affection, a loutish chef
at the restaurant where she works. A typical Hollywood version of
such an encounter—which takes place outside at night, in a metal
tube on a loading dock while both partners are mostly clothed—
would've been sleek and effortless, the woman instantly orgasmic.
In Dunham's hands, it went something like this: they kissed for
ten seconds; he unzipped his fly and wordlessly shoved her head
downward; he told her to "suck harder," cursed her incessantly
ringing phone, and then scuttled around her body to enter her
doggie style; he pounded into her until he ejaculated, which took
less than a minute; he never once looked at her face. Aura's expression shifted from aroused to confused, to slightly disappointed,
to resigned. Afterward, he bid her good-bye while checking his
texts. The scene is hard to watch without cringing—it's poignant,
it's agonizing, it's embarrassing, and it's real.

Young women grow up in a porn-saturated, image-centered,
commercialized culture in which "empowerment" is just a feeling, consumption trumps connection, "hot" is an imperative,

fame is the ultimate achievement, and the quickest way for a woman to get ahead is to serve up her body before someone else does. If Paris Hilton synthesized the zeitgeist ten years ago, it may be her former bestie, Kim Kardashian, who embodies it now. Kardashian is the Horatio Alger of the selfie set, pulling herself up by her bra straps and parlaying exhibitionism and a genius for self-promotion into an impressive eighty-five-million-dollar empire. Like Hilton, Kardashian came to prominence via a sex tape in a deal that, rumor has it, was brokered by her mother. She, too, seems strangely bored by the on-screen acts in that tape, chewing gum throughout. Still, the notoriety generated by the did-she-or-didn't-she-leak-it-on-purpose speculation was enough to pique the E! network's interest in a reality show. *Keeping Up with the Kardashians* (*KUWTK*) premiered in 2007. Soon after, Kim posed for *Playboy*, encouraged on *KUWTK* by her mother. By 2008 she was the world's most googled celebrity. Kardashian's personal brand would eventually extend to boutiques, fitness videos, clothing lines, skin care products, perfumes, a best-selling video game, and more. She wrangled eighteen million dollars in endorsements and broadcast rights for her 2011 wedding to pro basketball player Kris Humphries (the marriage lasted seventy-two days, prompting rumors that it had been a publicity stunt). By 2015 she was thirty-third on the *Forbes* list of the World's Highest Paid Celebrities. At this writing, she has more than forty-four million Instagram followers (she follows only ninety-six people herself) and has unseated Beyoncé as the site's most followed person. Kardashian reportedly earns up to twenty-five thousand dollars per sponsored tweet and an average of one hundred thousand dollars for personal appearances. Her aforementioned full moon on the cover of *Paper*, while it didn't "break the Internet," generated nearly sixteen million page views within thirty hours. She's now married to one of the foremost hip-hop artists on the planet—in an ode to their love, he penned a touching lyric about knowing she could be his "spouse, girl," "when I impregnated your mouth, girl"—and, together they have a daughter, North West, currently a toddler. I wonder how they'll react to *her* first sex tape.

Kardashian's A-list ascent has been a perfect storm of social media and pop and porn cultures, her celebrity not a result of talent, achievement, or skill but of the relentless pursuit of attention: she is famous for being #famous. Curiously, the adjective most used about Kim by her fans (besides *hot*) is *relatable*. She seems authentic to them even though they know that her "reality" is entirely artificial: staged, edited, curated, cross-promoted, co-branded, augmented, and enhanced. Perhaps more than anyone, she has mastered the body "product": figured out how, as a woman, to harness the contradictory demands of the media landscape and to do it for her own enormous profit. Again, this can be read as empowerment—if your definition includes perpetuating shopworn stereotypes about women. Girls point to her style, her work ethic, and her wealth—aren't those admirable qualities? Yet, as the blog *Sociological Images* pointed out after Kardashian's wax figure was installed in Madame Tussauds, Kim's true contribution has been an ingenious "patriarchal bargain": her acceptance of roles and rules that disadvantage women in exchange for whatever power she can wrest from them. It's difficult to see how Kardashian's success expands options for anyone but herself. (Okay, it helped her sisters.) It's feminism defined by "I've Got Mine," underscored by the winking title of her 2015 book, *Selfish*. Even Tina Brown, the former *Vanity Fair* editor who virtually invented high-low journalism, was concerned that a 2014

Vogue cover positioned Kim as "aspirational" to young women. Those aspirations, Brown wrote, "now have very little to do with any notion of excellence, either of character or of comportment. Our hopes have gotten so cheesy that even the cheese is ersatz."

If the script handed down by our hypersexualized culture expanded the vision of "sexy" to include a broad range of physical size and ability, skin shade, gender identity, sexual preference, age; if it taught girls that how their bodies feel to them is more important than how they look to others; if it reminded them that neither value nor "empowerment" are contingent on the size of their boobs, belly, or ass; if it emphasized that they are entitled to ethical, reciprocal, mutually pleasurable sexual encounters; then maybe, *maybe* I'd embrace it. The body as product, however, is not the same as the body as subject. Nor is learning to be sexually desirable the same as exploring your own desire: your wants, your needs, your capacity for joy, for passion, for intimacy, for ecstasy. It's not surprising that girls feel powerful when they feel "hot": it's presented to them over and over as a precondition for success in any realm. But the truth is that "hot" refracts sexuality through a dehumanized prism regardless of who is "in control." "Hot" demands that certain women project perpetual sexual availability while denying others any sexuality at all. "Hot" tells girls that appearing sexually confident is more important than possessing knowledge of their own bodies. Because of that, as often as not, that confidence that "hot" confers comes off with their clothes.

CHAPTER 2

Are We Having Fun Yet?

Alatte can be a great prop, kind of like a cigarette in a 1940s noir movie. Giving it a stir, taking a thoughtful sip, offers you time to gather yourself, which can be pretty vital when a virtual stranger, one who is basically old enough to be your mother, asks you point blank how often you masturbate, whether you've ever had an orgasm, or to describe your last sexual encounter with a partner. In fact, it gives the stranger asking the questions something to focus on as well, because, let me tell you, launching into a discussion of blow jobs with someone you've just met, someone young enough to be your daughter, can feel just a smidge uncomfortable. So I was relieved that Sam, eighteen, a senior at a California high school, had chosen to meet me on the patio of her favorite café, even if we were sitting next to a couple of middle-aged guys in Dockers and button-downs who were clearly shocked by our conversation. Sam was tall and full-figured, with golden skin and dark, loose curls that flowed nearly to the middle of her back. Her mother, a middle school math teacher, was African American; her father, whom she had rarely seen since her parents split, was white. Her mom had re-married a man from Samoa about five years ago; Sam calls him Dad. "I knew about romance and all that from an early age be-cause I'd meet my mom's boyfriends," Sam told me. "And when I went through puberty, she had books around."

I asked Sam whether her mother had explained to her about periods and reproduction. She nodded. What about masturba-tion? She laughed. "No," she said. The location of her clitoris? She laughed again. What about orgasm? She shook her head. "My parents are liberal," she said. "And they'll talk about sex gen-erally, or joke about it. We'll watch *South Park* or talk about the disfigurement of girls in the Middle East. But when it comes to *me*, it's a little more iffy. Then it's more like a conservative house-hold, where we don't talk directly. If I approached them, they'd be open to discussing it, but it's hard for them to bring up and it's hard for me to bring up."

Like most of the girls I met, Sam was both curious about sex and resourceful, so she did her own research on the subject, looking up on the Internet whatever she didn't know—through Google searches such as "how to give a blow job," or by checking out porn ("just to see how things fit together," she said). And, of course, she learned from doing. "Freshman year in high school was when everything became a reality," she recalled. "Sex, drink-ing, all of that. That's when you weren't just watching it on TV anymore. But we weren't *really* partying yet. It was mostly for ap-pearance. Like, you'd go to some park on the weekend and take a shot and sort of pretend you were drunk. And you'd hook up with some guy and maybe go to second or third base."

I stopped Sam right there. The terrain of relationships and sexual intimacy had changed since I was a girl. Along with it, there was a whole new vocabulary that both tripped me up and, as I'm someone passionate about words, fascinated me: *Talking*, for instance, did not mean conversation, but was a synonym for what, in an earlier era, would've been called "seeing" someone. As in "We're not serious, Mom. We're just talking." (It seemed a particularly ironic choice for today's teens, given their preference for connecting via text over actual conversation.) "Hooking up," a phrase that has inspired a full-scale media panic about the morals of a new generation, could mean anything from kissing to intercourse. Its ambiguity was the source of perpetual misunderstanding not only between girls and adults but among peers: *hooking up* was so vague a term that they could never be quite sure what their friends were up to. *Catching feelings* meant developing an emotional attachment and was, for many girls, something to protect against when hooking up, just as they would guard against catching herpes or chlamydia. A boy being "all cute" meant he may have "caught feelings" since he was behaving in a caring, thoughtful way toward a girl—what I would have called "romantic." *Dating*, though never a word much used beyond sixth grade, was the *last* step on the path toward a relationship, coming well after "hooking up" and "exclusive hooking up." Girls sometimes referred to their genitals as "my junk," and the phrase "making love" prompted gagging sounds. I couldn't help but notice that much of this new lexicon was devoid of terms not only connoting intimacy, but also indicating joy or pleasure.

So what, I asked Sam, was today's version of "the bases"?

She took a long swig of her latte. "Well, first base would be kissing," she said. "Second base would be a hand job for a guy and fingering the girl."

I raised my eyebrows. Already it seemed to me that a few steps had been skipped.

"And third base would be oral."

"Both ways?" I asked.

Sam laughed again and shook her head. "For the *guy*," she said. "Girls don't get oral sex. No. Not unless you're in a long-term relationship."

"Wait," I said. "Back up. I don't actually recall oral sex as being a base at all."

Sam shrugged. "That's a difference between my generation and yours," she said. "For us, oral sex is not a big deal. Everyone does it."

Why Do You Think They Call It a Blow "Job"?

There has been a lot of anxiety over the past couple of decades about teens and oral sex. Much of it can be traced back to the late 1990s, to a *New York Times* report that among middle-class teens, oral sex—and by "oral sex," it meant fellatio—not only was becoming ubiquitous, but that they were engaging in it far earlier and more casually than teens' busy (read: neglectful) working parents realized. One health educator was quoted as saying, " 'Do you spit or do you swallow?' is a typical seventh-grade question."

Two years later, the *Washington Post* covered a parent meeting called by middle school counselors in Arlington, Virginia, a town of "elegant brick homes, leafy sycamores and stone walls"—again, code for white and middle class—to discuss the fellatio craze among thirteen-year-old girls. The reporter linked that incident to a wider regional trend, based largely on "student grapevine"–generated claims of girls who had dropped to their knees during study hall or at the back of a school bus.

Girls' bodies have always been vectors for a society's larger trepidations about women's roles. It was likely no coincidence, then, that those early blow job scandals surfaced just as oral sex was making front-page news for another reason: the country was gripped by an obsession with a certain blue Gap frock and a cigar that was by no means just a cigar. President Bill Clinton's alleged dalliance with Monica Lewinsky, a White House intern less than half his age, dominated the headlines, sending mortified parents leaping from the couch to twist the radio dial or grab the TV remote when the latest reports aired. Most famously, in January 1998, Clinton testified under oath that "I did not have sex with that woman, Miss Lewinsky." A few months later, when DNA from his semen was discovered on the fabled dress that she had squirreled away as a memento of their tryst—and, might I say, *ick*—he insisted that he had not perjured himself because their relationship involved only *oral* sex. Suddenly, people across the nation were hotly debating whether mouth-to-genital contact was, indeed, "sex." If it wasn't, what exactly was it? And how were Americans supposed to explain the president's hairsplitting to their children?

Oral sex had only recently become a standard part of Americans' erotic repertoire. Historically, both fellatio and cunnilingus were considered *more* intimate than intercourse, acts to be engaged in only after marriage, if at all. In 1994, just a few years before the Clinton affair broke, *Sex in America*, the most definitive survey at that time to be released on this country's sexual practices, found that while only a minority of women over fifty had *ever* performed fellatio, among women under thirty-five, three quarters had done so. (Most men, whatever their age, said they had been both providers and recipients of oral sex.) The rise in going down among straight couples, the authors wrote, was the biggest sexual change of the twentieth century. By 2014 oral sex was so common as to be unremarkable: as one researcher quipped, the number of Americans who thought Barack Obama was Muslim was larger than those who had never given or received oral sex.

But the notion that the practice was aging downward, that among teens it was becoming more common and less meaningful than intercourse, was most definitely a new phenomenon, one that caught not only parents but also researchers off-guard. There was very little hard data to back those early journalistic claims. Oral sex practices of minors had been considered unfundable in academia; even if one could get the money, what parent would allow their child to be questioned on the subject? More generally, there was a presumption among conservative politicians that talking to teens about any form of sex, even in the name of research, was tantamount to handing them an instruction manual. Because of that, vital information about kids' sexual behaviors, including disease transmission, went virtually unstudied.

By 2000 the Clinton presidency was winding down, but the blow job panic had just begun. A new story in the *New York Times* declared that sixth-graders were now, basically, treating fellatio like a handshake with the mouth. According to one Long Island child psychologist, girls that age would tell him earnestly that they expected to wait until marriage for intercourse, yet had already given head *fifty or sixty times*. "It's like a goodnight kiss to them," he claimed, "how they say good-bye after a date." The director of the Parenting Institute at New York University, meanwhile, predicted that soon a "substantial" number of kids would be having intercourse by middle school. "It's already happening," he told the *Times*. (That was not true: according to the Centers for Disease Control and Prevention, in most states rates of intercourse among middle schoolers were dropping.) An article in the now-defunct *Talk* magazine blamed dual-career "parents who were afraid to parent" for an epidemic of oral sex among seventh-graders—again acting out larger anxieties about women, in this case working mothers, through concerns about unsupervised, wayward girls.

It was Oprah, however—isn't it always Oprah?—who sounded the loudest alarm. In 2003 she invited onto her show a reporter for *O Magazine* who had interviewed fifty girls about their sexual practices. "Hold on to your underwear for this one," the writer said, before revealing her ultimate stunner: the rainbow party. In this version of *Girls Gone Wild*, young women barely past their Barbie phase were donning different shades of lipstick, then fellating groups of boys in turns, leaving behind a "rainbow" of makeup on each penis. The girl whose color hit farthest down was declared the "winner."

Well, what parent wouldn't freak out? Children were having indiscriminate sex (or indiscriminate not-sex) everywhere! Under the table at bar mitzvahs! Behind the monkey bars during recess! No one, least of all Oprah, seemed to question the actual logistics of any of this. Exactly *how* were girls managing to complete multiple, random sex acts during the school day without an adult's notice? Were thirteen-year-old boys really up to fifteen public blow jobs in the space of a few hours? Wouldn't any rainbow effect be rinsed off or at least indelibly smudged by each subsequent partner? A 2004 NBC News/*People* survey taken shortly after the rainbow party story broke found that, in truth, less than one half of 1 percent of children ages thirteen to sixteen said they'd attended an oral sex party. Although that's not zero, it's hardly rampant.

So, no, children were not having orgies. That said, the seed from which the "rainbow party" myth sprouted did come from somewhere: oral sex has become relatively commonplace among teens. By the end of ninth grade, nearly one in five children has engaged in oral sex; by age eighteen, about two thirds have, with white and more affluent teens indulging more than others. Pinning that change on Bill Clinton or the sexual revolution or parental permissiveness, however, would be simplistic—and incorrect. Right-wing influence on sex education has played an equal, if not greater role. Federally mandated abstinence-only programs, which began in the early 1980s, not only reinforced that intercourse was the line in the sand of chastity, but also, using the threat of AIDS as justification, hammered home the idea that it might well kill you. Oral sex, then, was the obvious workaround. I doubt, though, that social conservatives would consider it a victory that, across a range of studies, college students who identify as religious are even more likely than others to say oral sex is not "sex," or that over a third of teenagers included it in their definition of "abstinence" (nearly a quarter included anal sex), or that roughly 70 percent agreed that someone who engages in oral sex is still a virgin.

I wondered, though: If teens didn't consider oral sex to be "sex," how did they perceive it? What did it mean to girls to give or receive oral sex? Did they enjoy it? Tolerate it? Expect it? One evening, shortly after her graduation from a suburban Chicago high school, a young woman named Ruby allowed me to join her and four of her friends for a chat. We met in Ruby's bedroom, one wall of which she'd painted midnight blue. Leggings, T-shirts, and skirts tumbled out of half-open dresser drawers. The girls sprawled on the floor, across the bed, on a beanbag chair.

When I asked about oral sex, a girl named Devon shook her head. "That's not a thing anymore," she said, waving a hand dismissively.

"So what is it, then?" I asked.

Devon shrugged. "It's nothing."

"Well, it's not that it's *nothing*," added Rachel.

"It's not *sex*," Devon countered.

"It's like a step past making out with someone," said Ruby. "It's a way of hooking up. A way to have gone farther without it being seen as any big deal."

"And it doesn't have the repercussions that vaginal sex does," Rachel added. "You're not losing your virginity, you can't get pregnant, you can't get STDs. So it's safer."

That, unfortunately, is not entirely true—though, again, because oral sex is ignored by parents and educators, there is a widespread belief among teens that it is risk free. The result is that while their rates of intercourse and pregnancy have dropped over the past thirty years, their rates of sexually transmitted diseases have not. Teens and young adults account for half of all new STD diagnoses annually and the majority among women. The new popularity of oral sex has been linked to rising rates of Type 1 herpes and gonorrhea (a disease that, about a decade ago, researchers thought was on the verge of eradication). Avoiding STDs, though, isn't really why girls engage in oral sex. The number one reason they do it, according to a study of high schoolers, is to improve their relationships. (Nearly a quarter of girls said this, compared to about 5 percent of boys.) What, though, did "improving a relationship" mean exactly, especially since so many also told me that oral sex, at least where fellatio was concerned, was a way to emotionally distance themselves from their partners, protect against the overinvestment they feared would come with intercourse. For years, psychologists have warned that girls learn to suppress their own feelings in order to avoid conflict, to preserve the peace in friendships and romantic partnerships. Was performing fellatio another version of that? Whether they hoped to attract a boy's interest, sustain it, or placate him, it seemed their partner's happiness was their main concern. Boys, incidentally, far and away, said that the number one reason they engaged in oral sex was for physical pleasure.

For both sexes, but particularly for girls, giving oral sex was also seen as a path to popularity. Intercourse could bring stigma, turn you into a "slut"; fellatio, at least under certain circumstances, conferred the right sort of reputation. "Oral sex is like money or some kind of currency," Sam explained. "It's how you make friends with the popular guys. And it's how you rack up points for hooking up with someone without actually having sex, so you can say, 'I hooked up with this person and that person,' and increase your social status. I guess it's more impersonal than sex, so people are like, 'It's not a big deal.'"

I may be of a different generation, but, frankly, it's hard for me to consider a penis in my mouth as "impersonal." Beyond that, I was concerned about the dynamics around oral sex: the morass of obligations, pressures, and judgments leveled at girls; the calculus and compromises they made to curry favor with boys while remaining emotionally, socially, and even physically "safe"; the lack of reciprocity or physical pleasure they described, or expected.

One afternoon in San Francisco's Golden Gate Park, I met Anna, a freshman at a small West Coast college. Anna had grown up in a politically liberal family and attended progressive private schools through twelfth grade. She wore skinny jeans with lace-up boots and had recently pierced the small flap of cartilage in front of her ear canal with a silver hoop; her long, wavy brown hair was swept to one side. "Sometimes," she told me, "a girl will give a guy a blow job at the end of the night because she doesn't want to have sex with him and he expects to be satisfied. So if I want him to leave and I don't want anything to happen . . ." She trailed off, leaving me to imagine the rest.

There was so much to unpack in that short statement: why a young man should *expect* to be sexually satisfied; why a girl not only isn't outraged, but considers it her obligation to comply; why she doesn't think a blow job constitutes "anything happening"; the pressure young women face in any personal relationship to put others' needs before their own; the potential justification of assault with a chaser of self-blame. "It goes back to girls feeling guilty," Anna said. "If you go to a guy's room and are hooking up with him, you feel bad leaving him without pleasing him in some way. But, you know, it's unfair. I don't think he feels badly for you."

In their research on high school girls and oral sex, April Burns, a professor of psychology at City University of New York, and her colleagues found that girls thought of fellatio kind of like homework: a chore to get done, a skill to master, one on which they expected to be evaluated, possibly publicly. As with schoolwork, they worried about failing or performing poorly—earning the equivalent of low marks. Although they took satisfaction in a task well done, the pleasure they described was never physical, never located in their own bodies. They were both dispassionate and nonpassionate about oral sex—socialized, the researchers concluded, to see themselves as "learners" in their encounters rather than "yearners."

The concern with pleasing, as opposed to pleasure, was pervasive among the girls I met, especially among high schoolers, who were just starting sexual experimentation. They often felt, for instance, that once they'd said yes to intercourse with a partner, they could never say no again, whether or not they were "in the mood." "I remember sort of hating it," said Lily, now a sophomore at a West Coast public university, about her sexual relationship with a high school boyfriend. "I wanted to please him, but it felt sometimes like we couldn't have a normal conversation because he was so distracted by wanting to have sex. And I couldn't really think of a *reason* to refuse"—not wanting to didn't seem adequate. "Sometimes I felt like I was just a receptacle for his hormones."

Those media-fueled sex panics tend to prey on parental fears about girls' promiscuity or victimization; the backlash dismisses both as overblown. Rarely does anyone ask the girls themselves what they think, what they gain from or enjoy about their experiences. Sam mentioned social status. Lily talked about pleasing a boyfriend. Gretchen, a seventeen-year-old classmate of Sam's, said she enjoyed the thrill, however short-lived, in having power over a boy. "I've gone down on four guys now. I don't even know, really, why I do it." She paused, chewing contemplatively on her lower lip. "I guess I like that feeling of 'Ha! You can't get this from anyone else. I am in control here!' You knew they *really, really* wanted it and you could be like, 'No! No!' and then they'd be like, 'Please! Please!' Because they were so desperate. That part's kind of fun. But it's definitely not the physical side of it, because that's so gross and it really hurts my throat. I mean, it's sort of fun getting in the rhythm of it. But it's never *fun* fun."

Performing oral sex could make girls feel like the more active partner in an encounter. By contrast, they described cunnilingus and intercourse as passive, like something that was being done *to* them, leaving them vulnerable. Those empowered feelings about fellatio, though, coexisted with their opposites: a lack of control, pressure to comply, the unspoken threat of danger. Sam commented that while her male peers had been warned not to coerce girls into intercourse, pushing for oral was fair game. Because of that, while she had "plenty of guy friends," she preferred not to be alone with them (which would, it seems, be an obstacle to true friendship). "In my social world, if you're hanging out alone with a guy, the usual expectation is that you're going to hook up with him," she explained. "And if you decide not to, he might try to pressure you. So I'll hang out with a guy at school, but I would *never* go to his house or to a movie or do anything that could be construed as more than 'just friends' unless I wanted that to happen. It's not that they'd force themselves on you; it's that there would be *pressure*. There would be *disappointment*. And there might be tension in our relationship if it didn't happen."

I want to be clear here. Sam was not a pushover, not a meek or mousy girl. She was an honor student, an editor on her school paper, a varsity tennis player. She identified as a feminist and casually bandied about terms such as *slut shaming*, *gender binary*, and *rape culture*. She was applying to top-tier colleges. She was an astute observer of her world. She was also, most definitely, immersed in it. Nearly all the girls I interviewed were bright, assertive, ambitious. If I had been interviewing them about their professional dreams or their attitudes toward leadership or their willingness to compete with boys in the classroom, I might have walked away inspired. A sophomore at an Ivy League college, a lacrosse player whose mother was a partner in a large law firm, bragged to me about the "strong women" in her family. "My grandmother is a firecracker at eighty-eight, and my mom is crazy, and my sister and I are going to be as crazy as they are," she said. "In my family, you have to have a personality and be loud. That's how we interact. It's like a form of feminine power and knowing yourself."

Even so, she described how, at age thirteen, she slipped into a bedroom with her best friend's older brother, a ninth-grader on whom she'd had a longtime crush. Although she had never kissed a boy, never held hands, never had a boyfriend, somehow—she doesn't remember the details—she ended up going down on him. Afterward, he never mentioned the incident again, so neither did she. Her subsequent sexual experiences, a handful of casual hookups, haven't been much different. "It's always the same unspoken sequence," she said. "You make out, then he feels you up, then you give him head, and that's it. I think girls aren't taught to express their wants. We're these docile creatures that just learn to please."

"Wait a minute," I countered. "Didn't you just tell me about all the strong women role models in your family, about how you were loud and have a big personality and didn't take shit?"

"I know," she said. "I think I didn't realize . . ." She paused, trying to reconcile the contradiction. "I guess no one ever told me that the strong female image also applies to sex."

Discussions of sexual assault and consistent, enthusiastic consent are, thankfully, becoming more common on college and some high school campuses, yet if teens think of fellatio as not-sex (or not "anything"), if it's thought of as an entitlement or considered an appeasement, then both girls' right to say no and boys' obligation to respect that are compromised, and the lines between consent and coercion and assault risk becoming blurred. "You know," Anna mused, "in some ways giving head is a bigger deal than sex. Because it doesn't necessarily do anything for *me*. So it's like doing the person a favor because you love and care about them. And if it's someone you're dating, there's an expectation that he'll reciprocate. But in hookups, guys are typically really douchey about it. And there's pressure for the girl to do it. So it's about how comfortable you are resisting that pressure or not. It gets awkward to keep resisting."

Most young men do, of course, take no for an answer. Yet nearly every girl I spoke with had at least one experience with a boy who had tried, despite her clear refusal, to coerce or force her into oral sex: verbally, via repeated texts, or by physically planting his hands on her shoulders and pushing downward. A sophomore at a midwestern public university, for instance, told me she felt lucky that she'd never been sexually assaulted. A few minutes later she described going back to a boy's room after a party during her freshman year. They kissed for a while, and then he attempted the shoulder push. She said no, and he backed off, only to try again a few minutes later, and then once more shortly after that. When she refused for the third time, he blew up. "Fuck you, then. I'll find someone else," he said and shoved her out of his room. It was the middle of the night in February and her dorm was two miles away. She cried the whole way home.

Another young woman, a freshman at a New England college, told me that she performed oral sex for the first time shortly after her sixteenth birthday. It was not by choice. "It was the summer after sophomore year in high school," she recalled. "I'd been talking to this guy for a while; he seemed nice. We were in the back of his car kissing. He just . . . I don't know how it happened. I was high, and that was confusing. He was very aggressive. He wanted to have sex, and I was like, 'I don't think this is a good idea.' He was not accepting. He kept trying for sex. And I was like *no*. So he sort of forced oral sex. He pushed my shoulders. And I didn't know how to get out of it. I was mostly just shocked. It wasn't a good feeling. And it's lasted. I never liked the idea of oral sex again after that. I still don't."

Girls have long been made the gatekeepers of male desire, charged with containing it, diverting it, controlling it. Providing reliable release from it had now become their responsibility as well. Oral sex had become their compromise, a loophole, a strategy for carrying out that expectation with the minimum of physical, social, or emotional fuss. "It's almost like . . . *clean*, you know what I mean?" a junior at a New York City public high school told me. I didn't know what she meant, not really. "It's like . . ." she said, "it's like . . . it's what's expected of you."

Girls rarely mentioned manually stimulating a boy. If the goal was to remain detached and impersonal, I would've thought that would be the obvious choice. "No," said Ruby in Chicago. "A guy can do that himself. 'A hand job is a man job. A blow job is yo' job.' Guys will actually say that. 'Just give me a blow job if you're going to do anything.'"

Listening to stories of obligatory, sometimes coerced, usually one-sided oral sex, I began to wonder: What if, rather than blow jobs, guys were expecting girls to, say, fetch them lattes from Starbucks? Would the girls be so compliant?

Sam laughed when I asked her that. "Well, a latte costs money . . ."

"Okay," I said. "Pretend it was free. Let's say guys expected you to keep getting them cups of water from the kitchen whenever you were alone. Would you be so willing? And would you mind that they never offered to bring you one in return?"

Sam laughed again. "Well, I guess when you put it that way . . ."

As Anna said, reciprocity in casual encounters was never assumed. That was fine with some girls, even a relief; those like Anna, however, who enjoyed oral sex, were miffed. "It's just expected that the guy will get off," she complained, "and then maybe he will be like"—she dropped her voice into a low register and gestured halfheartedly toward my torso with her chin—" 'Oh, uh, do you want me to . . . ?' It's never like he'll do something for me and *maybe* I'll do something for him. It's like, *naturally* I'll do something and then he'll ask if I 'want' him to."

One young woman I met, a college freshman who was a self-described "nympho" (who had also, she said, spent every summer of her teens at "Jesus camp") told me that she no longer tolerated lack of reciprocity from her "randoms." "The worst experience I had was when I hooked up with this guy and he got me down to my bra and underwear and he's in his boxers. Normally the next thing would be the bra comes off. But the bra didn't come off. Instead, all of a sudden his boxers were off. And then he did this"— she pantomimed the shoulder push. "And I was like, 'Wait, just because my organs are inside and yours are outside I'm not going to get *anything* and you expect me to go down on you?' I was like, 'We are done. This is not going to happen.' It was incredibly awkward, though. I had to get him out of my room."

It's Sacred Down There. Also Icky.

When my daughter was a baby I read somewhere that, while labeling their infants' body parts ("here's your nose," "here are your toes"), parents typically include a boy's genitals (at the very least, "here's your pee-pee") but not a girl's. Leaving something unnamed makes it quite literally unspeakable: a void, an absence, a taboo. Nor does that silence change much as girls get older. Adolescent penises insist on recognition. Enter any high school and you'll see them scrawled everywhere: on lockers, on notebooks, on desks, on clipboards. Boys cannot seem to restrain themselves from drawing their sexual organs, loud and proud, on any blank surface. But whither the bushy vulva, the magnificent minge, the triangular twat?

Did I hear an "eww"? Exactly.

Even the most comprehensive sex education classes stick with a woman's internal parts—uteri, tubes, ovaries. Those classic diagrams of a woman's reproductive system, the ones shaped like the head of a steer, blur into a gray Y between the legs, as if the vulva and the labia, let alone the clitoris, don't exist. Imagine not clueing a twelve-year-old boy into the existence of his penis! And whereas males' puberty is characterized by ejaculation, masturbation, and the emergence of a near-unstoppable sex drive, females' is defined by . . . periods. And the possibility of unwanted pregnancy. Where is the discussion of girls' sexual development? When do we talk to girls about desire and pleasure? When do we explain the miraculous nuances of their anatomy? When do we address exploration, self-knowledge? No wonder boys' physical needs seem inevitable to teens while girls' are, at best, optional.

Few of the heterosexual young women I interviewed had ever had an orgasm with a partner, though most, from time to time, had faked it, taking their cues from the soundtrack of porn videos. Around a third masturbated regularly, which was, I was surprised to discover, about average. About half said they had never masturbated at all. It's hard to imagine adults would stand for such ignorance or lack of curiosity about any other body part. Most girls waved away my questions on masturbation, saying things such as "I have a boyfriend to do that" (though these were the same girls who'd never come with a partner). Beyond making them dependent on someone else for their pleasure, this was yet another inversion of what they said boys believed: that since they could masturbate on their own (it was a "man job"), they didn't require a partner for that. As for being on the receiving end of oral sex, girls tended to describe allowing (let alone *wanting*) a boy to head south as an intimate, emotional act requiring a deep level of trust. "I was in a relationship with a kid for a year where I had given him head," recalled Rachel in Chicago, "but I never felt comfortable for him to return the favor. Because . . . okay, this is weird to say, but a guy going down on you is more like a sacred thing. Like once you've done *that*, you really must be comfortable with the person, because it is not something that I'm just going to let you do."

"I'd rather have sex before I did that," Devon agreed.

"A guy is totally aware of what he looks like down there," Rachel continued, "but I don't know what they're seeing on me. I can't see it."

"Well," I said, "there are these things called mirrors . . ."

"Yeah," Rachel said dryly. "I'm not going to do that."

It's understandable that girls wouldn't let a partner go "down there" if they themselves were squeamish about their genitals. They worried that their vaginas were ugly, rank, unappealing. Again, not new concerns—I recall hiding a can of FDS "feminine deodorant spray" in the back of a desk drawer in junior high— but how is it that they still persist? Hadn't these girls heard of *The Vagina Monologues*? Erin, a senior at a San Francisco high school who was president of her school's feminist issues club, boasted that she was "really good" at giving oral sex to her boyfriend of a year, but when I asked how she felt about receiving it, she wrinkled her nose. "He doesn't go down on me," she said. "He doesn't want to. And I've never asked. Because . . ." She took a deep breath. "I don't like my vagina," she admitted. "I know that sucks. And I don't know why it should be so different, but I've internalized that idea.

"It's like that whole thing about queefing," she continued.

"Queefing?" I asked.

"Yeah," she said. "It's a fart with your vagina? There were these episodes on *South Park* about it, and now teenage boys have that as something they can say about girls, and girls know that boys have that, so you feel awkward." She sighed. "It's just, there's this whole comedic culture around making fun of female sexuality, you know? And it's super strong."

While queefing had blessedly escaped my notice, the overall rise of the word *vagina* as a punch line had not. Snarky references to women's nethers are the new *fag*—a way to denigrate masculinity, to ridicule or dominate an opponent. Even women use the word to signal that they're "cool with it," down with the bros. The implication is that *everyone* shares a secret distaste toward a lady's parts, or at least a sense that the word *vagina* itself is a goof (as opposed to *cunt*, which wouldn't be funny at all, and *pussy*, which as an insult has lost much of its anatomic specificity). So in the 2007 film *Knocked Up,* Jason Segel taunts a bearded Martin Starr by saying, "Your face look like a vagina." In *Forgetting Sarah Marshall,* Mila Kunis similarly slams Segel, when he hesitates before diving off a Hawaiian cliff: "I can see your vagina from here," she calls from the ocean below. "I can see your hoo-hah." Another female character, in the trailer for the Adam Sandler flop *That's My Boy,* heckles an ineffectual Andy Samberg with "Throw [the ball] you big vagina!" Off-screen, an essay on the website Thought Catalog titled "I'm a Feminist, but I Don't Eat Pussy" went viral in 2013. Among its pithy observations: that while vaginas "feel really good when your penis is inside of them" they are "objectively gross . . . covered in hair. They ooze and slime . . ." They are dirty, the male writer continues, and taste bad, and for women to expect oral sex "when you know the strain it puts on men, is selfish and, frankly, discriminatory." If that weren't enough to plunge the average young woman into a shame spiral, heartthrob actor Robert Pattinson, whose fame and fortune were forged from the erotic fantasies of teenage girls, breezily confessed to *Details* magazine, "I really hate vaginas. I'm allergic to vagina."

Sign me up for Team Jacob.

No wonder girls are insecure. Remember the shoulder push? The wordless gesture boys use to urge their partners downward? Young women had their own version, but it was a two-palmed shove *away* from the pelvis, a silent redirection to safer, if less erogenous, ground. Sam said that her ex-boyfriend, whom she had dated for a year, went down on her exactly twice during their relationship. Both times it was his idea. "It was not fun for me," she said. "I was not comfortable with it at all. I guess because I've never been comfortable with my parts down there. It's not something I find attractive. So I don't like the idea of someone else down there." To be fair, she said, he would "finger" her, but he had no idea what felt good; nor, since she had never masturbated, did she: even if she did know, she probably couldn't have said it out loud. Mostly, he just inserted a finger and sort of rummaged around.

Obviously, I wouldn't expect girls to be fully aware of their sexual needs or able to articulate them easily—many adult women can't do that even with long-term partners—but they are at a critical juncture in their development, learning foundational lessons about attraction, intimacy, arousal, sexual entitlement. Those early experiences can have a lasting impact on the understanding and enjoyment of their sexuality. So their aversion to their own genitals was disheartening. Watching girls squirm in response to my questions, I thought again about the images of female sexiness that assaulted them: Fergie's "London Bridge" going down, Miley swinging naked on a wrecking ball, Beyoncé dancing in her scanties around her suit-clad husband, Nicki giving Drake a lap dance (tweeting beforehand that she had just

GIRLS & SEX

knocked back some "confidence juice"). The culture is littered
with female body parts, with clothes and posturing that purport-
edly express sexual confidence. But who cares how "proud" you
are of your body's appearance if you don't enjoy its responses?
One sophomore in college showed me photos from her Instagram feed in which she was dressed for a party in a leopard-print
crop top, a tiny skirt, and skyscraper heels. Later in our interview
she admitted, "I don't enjoy getting oral sex. I am so in my head.
All I think about is if I should tell him that it doesn't feel good or
if he's getting tired or if he's even grossed out?"

Women's feelings about their genitals have been directly linked
to their enjoyment of sex. College women in one study who were
uncomfortable with their genitalia were not only less sexually satisfied and had fewer orgasms than others but were more likely to
engage in risky behavior. (Boys were the opposite: those who felt
positively about their penises were more likely to engage in risky
sexual behavior.) Another study, of more than four hundred undergraduates, found that early engagement in fellatio led to feelings of inferiority and low self-worth among girls; by contrast,
cunnilingus at the same young age was associated with greater self-
awareness, sexual openness, and assertiveness. Young women who
feel confident masturbating during sex, meanwhile, more than
double their odds of orgasm in either hookups or relationships.

So how young girls feel about "down there" matters. It matters
a lot.

The Psychological Clitoridectomy

Sex is probably not the first thing that jumps to mind when you
think about Indiana. But it happens that the state university in

66

Bloomington is home to the Kinsey Institute, a center of research on sexual health founded by biologist Alfred Kinsey. I flew there one icy winter afternoon to meet Debby Herbenick, an associate professor at IU's School of Public Health. Herbenick, who is also a sex columnist and the author of books such as *Sex Made Easy*, was the very picture of the modern sexpert: in her late thirties, with long, dark hair and cocker spaniel eyes, and dressed in a chic houndstooth minidress with over-the-knee boots. Her own research is in an area called genital self-image: how people feel about their private parts. Over the past few years, she said, young women's genital self-image has been under siege, with more pressure on them than ever to see their vulvas as unacceptable in their natural state: "They need to shave them, decorate them, or otherwise groom before sex," she said. "There's this real sense of shame as a girl if you don't have your genitals prepared, a real sense that there is a possibility someone will judge them."

Most of the young women I met had shaved or waxed their pubic hair, all of it, since they were about fourteen. When I asked them why, the girls would initially say it wasn't something they'd ever questioned: they already shaved their legs and armpits, and they'd seen older girls who were bare, so it seemed the thing to do. They said hairlessness made them feel "cleaner" (mistakenly, as it turns out. Though it diminishes the risk of pubic lice, clear cutting creates a festive-sounding "happy culture" for most other STDs: without the shield of protective hair, for instance, the labia can become carpeted with genital warts). As with self-objectification, girls considered depilation a personal choice, something done "for oneself," for comfort, hygiene, practicality. Invariably, though, they would bring up another motivation: avoiding humiliation. Consider the trajectory of comments by Alexis, a sixteen-year-old at a public high school in Northern

California. "I didn't really think about it," she began. "One friend had an older sister who was doing it, so she started, and then we all did it. It was like a chain reaction.

"But then, I also heard these guys in class one day talking about a girl. Her shorts were low cut and when she'd raised her arms, her shirt had lifted up, and they were like, 'I could see pubic hair! Man, it was so gross!'"

Girls are already self-conscious about their (typically unnamed) pubic region; it doesn't take much to stoke that insecurity. Ruby, in Chicago, was one of the girls who said shaving made her feel "clean," especially during her period. But she, too, added, "I remember these boys telling stories about this girl who 'got around.' And guys would go down there to finger her, or whatever, and there would be hair, and they were appalled. So I just . . . I mean, guys act like they would be *disgusted* by it."

Herbenick said that in her college town, chalkboard displays outside local salons offered sales on "back-to-school waxes" in the fall; April brought similar specials on spring break Brazilians. "That's a pretty public reminder that you better look a certain way," she said. A few years ago, she had a female student confide that she'd started shaving after a boy announced—during one of Herbenick's class discussions—that he'd never seen pubic hair on a woman in real life, and if he came across it on a hookup partner, he'd walk out the door.

Full-frontal waxing—which is not only pricey but excruciating—was once the province of fetishists and, of course, porn stars. The first "Brazilian" salon (so named because its owners were from that country) in the United States opened in New York in 1987, but it was an episode of *Sex and the City* that took the practice mainstream. By 2006, trendsetter and the former Posh Spice Victoria Beckham declared that Brazilian waxes should be "compulsory" starting at fifteen. (Let's check back with her in 2026, shall we, when her daughter reaches that age.) There's no question that a bald vulva is smooth. Silky smooth. Baby smooth—some would say disturbingly so. Perhaps in the 1920s, when women first started shaving their legs and armpits, that, too, seemed creepily infantilizing, but now depilating those areas is a standard rite of passage for girls, an announcement, rather than a denial, of adult sexuality. That first wave of hair removal was driven by flapper fashions that displayed a woman's limbs; arms and legs were, for the first time, no longer part of the private realm. Today's pubic hair removal may indicate something similar: we have opened our most intimate parts to unprecedented scrutiny, evaluation, commodification. Largely as a result of the Brazilian trend, cosmetic labiaplasty, the clipping of the folds of skin surrounding the vulva, has skyrocketed: while still well behind nose and boob jobs, according to the American Society of Aesthetic Plastic Surgeons (ASAPS), there was a 44 percent rise in the procedure between 2012 and 2013—and a 64 percent jump the previous year. Labiaplasty is almost never related to sexual function or pleasure; it can actually impede both. Never mind: Dr. Michael Edwards, the ASAPS president in 2013, hailed the uptick as part of "an ever-evolving concept of beauty and self-confidence." The most sought-after look, incidentally, is called—are you ready?—the Barbie: a "'clamshell'-type effect in which the outer labia appear fused, with no labia minora protruding." I trust I don't need to remind the reader that Barbie is (a) made of plastic and (b) *has no vagina.*

Herbenick invited me to sit in on the Human Sexuality class she was about to teach, one of the most popular courses on Indiana's campus. She was, on that day, delivering a lecture on gender disparities in sexual satisfaction. More than one hundred fifty students were already seated in the classroom when we arrived, nearly all of them female, most dressed in sweats, their hair pulled into haphazard ponytails. They listened raptly as Herbenick explained the vastly different language young men and young women use when describing "good sex." "Men are more likely to talk about pleasure, about orgasm," Herbenick said. "Women talk more about absence of pain. Thirty percent of female college students say they experience pain during their sexual encounters as opposed to five percent of men."

The rates of pain among women, she added, shoot up to 70 percent when anal sex is included. Until recently, anal sex was a relatively rare practice among young adults. But as it's become disproportionately common in porn—and the big payoff in R-rated fare such as *Kingsman* and *The To Do List*—it's also on the rise in real life. In 1992 only 16 percent of women aged eighteen to twenty-four said they had tried anal sex. Today 20 percent of women eighteen to nineteen have, and by ages twenty to twenty-four it's up to 40 percent. A 2014 study of heterosexuals sixteen to eighteen years old—and can we pause for a moment to consider just how young that is?—found that it was mainly boys who pushed for "fifth base," approaching it less as a form of intimacy with a partner (who they assumed would both need to be and *could* be coerced into it) than a competition with other boys. Girls were expected to endure the act, which they consistently reported as painful. Both sexes blamed that discomfort on the girls themselves, for being "naïve or flawed," unable to "relax." Deborah Tolman has bluntly called anal "the new oral." "Since all girls are now presumed to have oral sex in their repertoire," she said, "anal sex is becoming the new 'Will she do it or not?' behavior, the new 'Prove you love me.'" And still, she added, "girls' sexual pleasure is not part of the equation." According to Herbenick, the rise of anal sex places new pressures on young women to perform or else be labeled a prude. "It's a metaphor, a symbol in one concrete behavior for the lack of education about sex, the normalization of female pain, and the way what had once been stigmatized has, over the course of a decade, become expected. If you don't want to do it you're suddenly not good enough, you're frigid, you're missing out, you're not exploring your sexuality, you're not adventurous."

I recalled a conversation I'd had with Lily, the girl who was exasperated by her high school boyfriend's preoccupation with intercourse. He watched a lot of porn, too, she'd said, and was particularly game to try anal sex. She complied mostly because she wanted to please him. "The first time, we had to stop right away because I hated it," she said. "Later, he pressured me to do it again; he said that we hadn't actually done it before, since it was so short. At that point I guess I did it out of stubbornness. Like, *Okay*, fine. *I'll do it again and I still won't like it*." She laughed. "Which clearly isn't very healthy."

In sexual encounters, girls, it seemed, were growing more accustomed to coercion and discomfort than, say, orgasm, afraid to say "no" lest they seem uptight. Consider that at every age three-quarters of men report regularly climaxing during partnered sex, while only about 29 percent of women do. Or that girls are four times more willing than boys to engage in sexual activity they don't like or want, particularly oral and anal sex. Women also, Herbenick said, use more negative language than men when describing *unsatisfying* sexual experiences. Again, they talked about pain. But they also talked about feeling degraded and depressed. Not a single man surveyed expressed similar feelings. According to Sara McClelland, who coined the term *intimate justice*, the whole notion of comparing women's and men's reports of "sexual satisfaction" assumes a common understanding of what the phrase means. Clearly, that's not the case. Not if, going into their encounters, women anticipate less pleasure and more pain than men. Among the college students McClelland studied, women tended to use their *partner's* physical pleasure as the yardstick for *their* satisfaction, saying things such as "If he's satisfied, then I'm sexually satisfied." For men, it was the opposite: the measure was their own orgasm. (Women's commitment to their partners' satisfaction, by the way, was independent of that person's gender, which may explain, in part, why girls are more likely to orgasm in same-sex encounters.) So when young women report sexual satisfaction levels equal to or greater than young men's—which they often do in research—that may be deceptive. If a girl goes into an encounter hoping it won't hurt, wanting to feel close to her partner, and expecting that *he* will have an orgasm, then she'll be satisfied if those criteria are met. There is nothing wrong with wanting to feel close to a partner or wanting him to be happy, but "absence of pain" is a pretty low bar for your own physical fulfillment. As an eighteen-year-old high school senior told me, "I understood before I started having sex what it meant for a guy to finish. You know it has to happen for sex to be over and for them to feel good. But I had no idea what it meant for a girl. Honestly? I still don't know. It's never addressed. So I've gone into it all without really understanding myself."

Listening to girls' litany of disembodied early experiences, it sometimes struck me that we'd performed the psychological equivalent of a clitoridectomy on our daughters: as if we believed, somehow, that by hiding the truth from them (that sex, including oral sex and masturbation, can and should feel fabulous) they won't find out, and so will stay "pure." What if the opposite were true: what if understanding one's physical responses, truly "expressing your sexuality" instead of just impersonating sexiness, could actually raise girls' expectations of intimate encounters? What if self-knowledge encouraged them to hold a higher standard for their experiences, both within and outside relationships? What does, or should, "sexually active" mean, anyway? Clearly, the classic definition is obsolete. It may be that we have to reconceptualize "sex" entirely, starting with virginity.

CHAPTER 3

Like a Virgin, Whatever That Is

Christina Navarro sat cross-legged on a pillow on the floor of her college cooperative, watching a YouTube video on her laptop. On the screen a forty-something woman named Pam Stenzel paced back and forth in front of a sign that read, "The High Cost of Free Love." She was dressed in a denim jacket and jeans, pontificating in a gravelly, I'm-down-with-the-kids voice about virginity. "If you're here today and you're a virgin," she said to a rapt audience of high schoolers, "*Good for you! GOOD FOR YOU!* You have something *so* special, *so* valuable it is worth *whatever it takes* to get to your marriage with no past, fear, or disease." The students cheered and applauded.

Stenzel is one of the nation's most renowned (or, depending on how you look at it, most notorious) abstinence-only educators, invited to the White House and the United Nations, a guest on programs such as *The Dr. Laura Show* and *Politically Incorrect with Bill Maher.* Allegedly the daughter of a rape victim who was adopted into a Christian home, she has dedicated her life to promoting chastity and exalting virgins. She earns as much as

$5,000 a gig; according to tax records, her company, Enlighten Communications, takes in around $240,000 a year.

I watched Christina watch the video, a look of amusement playing across her face. She was twenty, though looked and sounded several years younger, and a junior at a public university on the West Coast. The walls of the room in which we sat were painted a deep purple; Indian print bedspreads were tacked to the ceilings and covered the mattress. A plate with the remains of a vegetarian burrito lay on the floor by the door. If I didn't know better, I could've sworn I'd time-tripped back to 1980. Just last week, Christina told me, the residents of the house had engaged in a spirited debate (one that brought up nostalgia for my own college days) over whether women should be free to walk around topless in the common areas. "It sparked this long conversation about how women's breasts have been objectified and sexualized by the media," she said, "and how in our house we should be able to express our bodies and be safe." Naturally, the decision was ultimately determined by consensus.

Earnestly naked girls in cooperative student housing may seem a long way from Pam Stenzel's chastity rants, but Christina grew up in Colorado Springs, Colorado, one of the most conservative cities in America and home to so many fundamentalist Christian organizations that it has been dubbed the Evangelical Vatican. Christina herself was not raised in that tradition—she's Catholic—but the "sex education" she received at her small parochial school was essentially identical, encapsulated by one word: *don't*. Rather than a Health Class, human sexuality was covered in tenth-grade Theology. The curriculum consisted mostly of scary statistics about the inevitability of pregnancy and disease for those who engaged in premarital sex and of the perils of abortion. Students were directed to memorize Bible passages interpreted as condemning homosexuality and advocating chastity. Watching Stenzel's videos in that class was an annual event, Christina recalled, a kind of rite of passage among her classmates, similar to the way watching gruesome films of incinerated accident victims once was for those who took Driver's Ed. Stenzel, who is based about an hour away from Colorado Springs, even lectured in person at an assembly at Christina's school. She was greeted with the anticipation and hoopla of a rock star. Even at the time, Christina said, she suspected Stenzel's presentation was "biased," and a little cheesy, but she didn't necessarily consider it inaccurate. And she never questioned the value of remaining "pure" until marriage.

On-screen, Stenzel was still talking: "Once it's gone, it's gone," she warned. "It takes that long"—she snapped her fingers—"to throw it away. It takes a lot of integrity to wait."

There was more applause, and then the video ended. We were quiet for a moment. "Are you still planning to save yourself for marriage?" I asked Christina.

She laughed and shook her head. "Oh, no," she said. "It's too late for that."

Cashing in the "V Card"

Nearly two thirds of teenagers have intercourse at least once before college—the average age of virginity loss in this country, as I've said, is seventeen—and while most do so with a romantic partner, a sizable number of girls cash in what they call their V card with a friend or a guy they've only just met. Over half, both in national samples and among my own interviews, were drunk for the occasion. Most say they regret their experience and wish they'd waited—maybe not until marriage, but longer than they did.

In some ways, I was surprised that the girls I talked to still considered first intercourse such a milestone. Most of them had already been sexually active for several years at that point, but again, that's assuming you "count" oral sex (or anything other than intercourse) as "sexually active." One could argue that in the modern world, "virginity" as a symbol of sexual initiation is an outdated, meaningless concept. It never had actual medical basis anyway (many girls have no hymen or have torn it through exercise, masturbation, or with a tampon), nor even a fully agreed-upon social meaning: in her book *The Purity Myth*, for instance, Jessica Valenti writes about the notion of "secondary virginity," the idea that virginity can magically be reinstated even after intercourse if one subsequently commits to abstinence until marriage. While that allows purity advocates to embrace those who have "stumbled," it also shows how arbitrary the definition of "virginity" can be. I'm not suggesting that first intercourse is psychologically or physically insignificant. Not at all. But why do girls in particular still elevate this single act (which, among other things, is rarely initially pleasurable for them) to a status beyond all others? Why do they imagine this one form of sexual expression will be transformational, the magic line between innocence and experience, naïveté and knowing? How does this notion of "virginity" as a special category shape their sexual experience? How is it affecting their sexual development, their self-understanding, their enjoyment of sex, their physical and emotional communication with a partner?

On a mellow fall Sunday morning, I joined Christina again, this time with a group of her friends on the rooftop veranda of the co-op. The other girls listened wide-eyed as Christina talked about her background; they found her stories exotic and a little shocking. "It's so surprising to me," said Caitlin, pushing her purple Clark Kent glasses up the bridge of her pierced nose. "At my high school they gave out condoms for free. They handed out *lube!*"

Even Annie, a freckled girl who attended an all-girls Catholic school in Orange County, California, considered her upbringing to be liberal compared to Christina's. "In high school my teacher unwrapped a peppermint patty and put it on the floor," Annie recalled. "Then she asked if we would eat it. Of course we were all, 'Eww, no!' And she said, '*Exactly!* Once you're "open," nobody will want you!' "

The girls cracked up. "But, then," Annie added, "my mom was kind of a hippie. So she would tell me to forget all that. She'd say, 'It's really important to test-drive a car before you buy it; you don't just kick the wheels.' "

When Brooke was in middle school, her mother gave her a pile of old-school sex-positive books such as *Our Bodies, Ourselves.* ("They all had these totally seventies covers," she recalled. "It was hilarious!") As for Caitlin, whose public high school passed out free condoms, when she was fifteen her mom took her to a "woman-friendly" sex store to buy a vibrator. "She said, 'I think it's really important that you get in touch with your own body and sexuality before you start having sex with someone else.' "

Neither Caitlin nor Brooke ever imagined saving her virginity for marriage. Until meeting Christina, they'd never met anyone who'd even considered it. "I think my mother's exact words were 'Virginity is a patriarchal construct,' " Caitlin said, and laughed. She had intercourse for the first time at sixteen, with a boy whom she would date for the next three years. "I would have actually done it earlier, with a different guy, my sophomore year," she said, "but he never initiated it. And I'm glad. Because I would have. Not because I *wanted* to have sex with him, but because I wanted to please him and I wanted to feel important. When I finally did have sex, it was only two months into my relationship, but I felt like I *wanted* to. It was really empowering to be absolutely sure of that decision and to realize that I hadn't been ready before but now I definitely was."

Brooke's first intercourse was younger still, at fifteen. She had imagined it would happen with a boy she cared about—she never used the word *love*—in the kind of romantic, gauzy setting you'd find in a vintage Summer's Eve douche commercial: on the edge of a cliff with the Pacific Ocean crashing against the rocks below. "I was probably thinking more about what it would be like to remember it later than the act itself," she admitted. "Like, how it would sound as a story."

It didn't go quite that way: Neither Brooke nor her boyfriend of seven months had a car, for one thing, so there was no way to get to the beach. Plus, it was winter. Anyway, what if someone walked by and caught them? In the end, they lost their mutual virginity in a fairly mundane fashion: in his bunk bed during a weekend when his family was out of town. She brought the condoms, which she had spent ages choosing at a nearby Walgreens, and the lube; she also, for reasons she can't remember, brought over a batch of home-baked cookies. "The truth is, losing your virginity is about the least sexy sexual act there is," she said. "It's awkward, especially when losing it to another virgin. Putting on the condom is the opposite of smooth. Things don't seem like they're going to fit together. You don't know how much of your weight to put on the other person. It's a little sweaty. And it doesn't feel good." After a minute or so, they felt like they had "done it" enough to say they had (both to themselves and to their friends), so they just . . . stopped. "But, you know," Brooke added, "it was a very positive experience for me. We bonded over the awkwardness, and that was fun. And even though the sex was lackluster, I felt totally comfortable with the situation and with him, and I'm grateful for that." They slept together a few more times before breaking up; Brooke kept their first condom wrapper as a souvenir, inscribing it with the date she used it.

Both Brooke and Caitlin were relieved to have lost their virginity when and how they did: too many of their friends, they said, were panicked about unloading "it" before college and, as a result, had made hasty choices that led to unpleasant experiences. College had loomed as a deadline for most of the girls I spoke with: being tagged as a prude freshman year seemed a greater threat to them than being labeled a slut. Better to get it over with, have sex with *someone*, rather than risk being seen as an "inexperienced freak" or, worse, as "too ugly to fuck." In general, young people overestimate how many of their peers have had sex, how many times they've had sex, and how many partners they've had (not to mention whether any of that sex has felt good). One in four eighteen-year-olds hasn't had intercourse. However, unless they're religious, most don't advertise their status—some even lie about it. Christina, who as a college freshman still expected to remain abstinent until marriage, felt she had to perpetually defend that choice, putting it out there right away when she met a guy at a party, to avoid any pressure or assumption. "But if you think it through," said Brooke, "it's ridiculous what happens. I mean, you're seventeen, you're graduating high school, and you're so worried about going to college a virgin that you get drunk and have sex with some random guy. It's not like that prepares you for anything. It's not like it gives you all this experience or understanding of sex. People, myself included, talk like just doing the act changes you . . ."

"*Oh my God!*" Annie broke in. She'd had intercourse the first time last year, at age nineteen, with her longtime boyfriend. "I thought it would be this whole new world after I had sex the first time! I had learned at school and at church that when you find the 'right person' and you're really in love and you have sex, you will be transformed. Like, this veil would be lifted. But I didn't feel that way. I didn't feel like a new person. There were no birds chirping or bells ringing. And I thought, 'Oh my gosh, maybe it wasn't the right time after all, or maybe we didn't do it right.' I felt like I'd been sold a bill of goods."

In her book *Virginity Lost*, Laura Carpenter finds four ways young people relate to virginity, each, more or less, reflected in what these girls described to me. The first group believed virginity was like a gift: a precious expression of love, though one no longer connected to marriage. Like Annie, and to a degree like Brooke, "gifters" romanticized their first time—the person, the setting, the significance—wanting everything to be "perfect" and expecting sex to strengthen their relationships, deepen a partner's commitment. If the experience wasn't up to snuff, especially if they felt tricked or coerced into intercourse, they were devastated. Worse yet, the betrayal often left them feeling worthless, unable to assert themselves in future relationships. "Having given away her virginity to someone who clearly didn't appreciate it," Carpenter wrote of one such girl, "Julie felt diminished in value, so much so that she believed she was no longer special enough to refuse sex with less special men." The risk for "gifters," then, according to Carpenter, was that their experience, of virginity loss itself and beyond, was defined by their partners' reactions.

At the other end of the spectrum were those who treated virginity as a stigma, viewing it with mounting embarrassment and dismay as they neared high school graduation. They imagined first intercourse would be kind of like a reality show makeover, changing them instantly from duckling to swan, from child to adult. Relationships? Romance? Forget it. This group just wanted it *out of the way*. Although they tended to be more satisfied with the experience than those who saw virginity as a gift of love (largely because their expectations were so much lower), they were often disillusioned by how little actually changed for them in the aftermath.

Nearly a third of Carpenter's subjects, similar to Caitlin, saw virginity loss more as a process, a rite of passage: part of, but not the determining factor in, becoming an adult. They neither idealized virginity nor saw it as a burden; first intercourse was just a natural, inevitable step in growing up and exploring sexuality. They felt more in control of their choices than the other groups—especially over with whom they had sex and when. They also tended to have experimented extensively with at least one other partner before intercourse, and saw doing "everything but" as worthy in its own right. By contrast, those who considered virginity a gift saw "lesser" sex acts mostly as a way to measure their partners' trust and commitment; those who saw virginity as a millstone considered anything short of intercourse to be a letdown, a consolation prize.

Like most Americans today, the young people in these three groups did not expect to remain abstinent until marriage. At the same time, Carpenter found that a substantial minority of teens, which once would have included Christina, had gone resolutely the other way, becoming more committed to and more vocal about remaining chaste until their wedding night. For them, too, virginity was a "gift" to be shared with one true partner, but it was also something else: a way to honor God.

Waiting for the Prince

An attractive couple stepped out of a low-slung sports car at the entrance of the East Ridge Country Club in Shreveport, Louisiana. He dark-haired, in a tuxedo; she in what appeared to be a wedding dress: strapless, with a sparkling white bodice and yards of floor-length tulle. At second glance, though, I saw that something did not quite fit: there was a touch of gray at the man's temples. The woman was not actually a woman at all: she was a fourteen-year-old girl. These were not newlyweds; they were father and daughter, here for the seventh annual, tristate, Ark-La-Tex Purity Ball. Inside, other couples, similarly dressed, milled around a table laden with candy: pink and orange jelly beans and gum balls. Most of them were white, though there was a smattering of African Americans and a few Latinos. One group of daughters and dads (or other male "mentors," who were equally welcome) stood near curtains covered in twinkle lights. Some had already taken their seats at round tables decorated with candles and silk flowers. A few posed for commemorative photos of the evening, which, according to its online invitation, was "designed to equip and encourage young women seventh through twelfth grade to stay pure until marriage." For one hundred dollars a couple (plus fifty dollars each for any additional daughters), it continued, "this event allows fathers an opportunity to pledge themselves to love and protect their daughters. It also helps young women begin to realize the truth: that they are infinitely valuable princesses who are *worth waiting for*'" (emphasis in original).

The world's first Purity Ball was organized in 1998, in Colorado Springs, the town where Christina grew up, by a pastor named Randy Wilson. As the father of seven children, five of them girls, Wilson believed it was his duty to "protect" his daughters' virginity. It's unclear how many such events are held each year—for a while, reports placed the number at fourteen hundred internationally, but that turned out to be hype. More accurate figures are hard to come by, especially because the balls, like any community event, wax and wane based on local interest and organizers' skills. Either way, they are an outgrowth of a larger "True Love Waits" movement launched by the Southern Baptist Convention in the mid-1990s. The first year of the campaign, more than 100,000 young people joined, signing a pledge to remain abstinent until marriage. By 2004 more than 2.5 million had pledged—1 in 6 American girls. Another campaign, Silver Ring Thing, which was until 2005 partially funded by the U.S. government, has held more than 1,000 events, using Christian rock, hip-hop, and a high-energy, club-like atmosphere as a draw; more than 50 were scheduled in the first half of 2015.

The ball I attended was somewhat unusual, in that although it focused on fathers and daughters, it was organized entirely by women. Its founder, Deb Brittan, was, when it began, a sexual health educator at a local Crisis Pregnancy Center, the kind of organization that steers women with unplanned pregnancies away from abortion and toward adoption or parenting. "My heart was and has always been that I want those baby girls to have the best sex," she told me as the partygoers tucked into a dinner of baked chicken breast and potatoes. "Obviously, when you look at statistics, it becomes *very evident* that the healthiest choice and the only guarantee of not getting a sexually transmitted disease or becoming three times more likely to attempt suicide is to look at a commitment to abstinence until marriage."

I made a note to myself to check that suicide figure: it wasn't wrong; it came from a 2003 study by the conservative Heritage Foundation. The link between sex and suicide, though, could hardly be called causal. Girls, for instance, are also more likely than boys to be bullied and stigmatized for sexual activity, which in itself puts them at risk of depression and suicide. So it may be the shaming of sexually active teens rather than sex itself that is the problem. It may also be that teens who are already depressed are more likely to engage in and subsequently regret sexual activity. Or it may be that teens' expectations of sex are media-driven and unrealistic; or that having first intercourse specifically while drunk puts a child at greater risk. Whatever the case, Brittan's job was to go into local public and private schools and, like Pam Stenzel, give students her version of the facts of life. "Whatever they choose after that is up to them," she told me. "But by the time I got done"—she winked and gave me a playful nudge with her elbow—"they couldn't say no one told them."

EARLIER THAT DAY, I had stopped by the country club to chat with some of the previous ball attendees, who return each year to help with the festivities. Several of them wore sweatshirts bearing the initials S.W.A.T., for "Sisters Walking Accountable Together," a club formed to support girls in vows of chastity. I had changed my own shirt three times before heading out the door. A loose-fitting scoop-necked sweater over a tank top suddenly seemed too revealing, especially since the sweater tended to slide around and show a bra strap. A cardigan over the same tank also seemed potentially immodest. I settled on a boat neck pullover, hoping it wouldn't be seen as too tight. These are not, I hasten to say, my typical thoughts as I dress in the morning, but somehow the Purity Ball's emphasis on "modesty" and "purity" made me feel more conscious of how my body and self-presentation might be judged by others than I had been since I was a teen.

In the daylight, the ballroom was relentlessly beige, its view a winter-drab golf course under a sodden sky. Several girls were tying pink tulle bows to chairs. Haylee, a high school senior, dressed in sweatpants and a S.W.A.T. sweatshirt, stood back to survey the effect, hands on her hips, brow furrowed. "I think it might be a little too 'Sweet Sixteen,'" she said.

"But that's how old the girls are!" another girl countered.

Haylee was, in many ways, like any of the girls I met: she was bright and articulate, excelled in school, was athletic. (She'd played soccer competitively since she was five years old and taught windsurfing in the summer.) She even attended what she called a "hipster, do what you want" liberal arts magnet school. On a chilly Saturday, she had wound her hair into a messy bun, and the red polish on her short nails was chipped. I asked if there were many other pledgers among her classmates. She snorted with laughter. "No," she said. "Actually it's very easy to be anything at my high school *except* for a Christian. People are very accepting of whatever gender you think you want to be. That's cool. And you can be whatever sexuality you want to be, too, *except* for pure. It's strange that way. Most people, when I talk about the Purity Ball are like, 'You are so judgmental.' And I say, 'You're being judgmental about me!'" As a result, she said, she has few friends at school, mostly hanging around a small group of like-minded athletes. For her, she said, the ball, which she first attended four years earlier, at age thirteen, had been a revelation. "I'd never really felt special the way the ball made me feel," she explained. "I didn't know I could love and be loved the way I can and do now."

Haylee had never had a boyfriend. "My school is two-thirds girls, and most of the guys are gay," she said cheerfully. If she did, though, she thinks she would draw the line at hand-holding.

Possibly a kiss, but nothing more. "I think it would be really cool not to have your first kiss until your wedding day," she said. Other girls around us agreed. One said she would never be alone in a room with a boy, not even a dark movie theater—"*Maybe* a two-person table at a restaurant," she allowed. Another girl limited hugs with her boyfriend to three seconds so, she said, "things wouldn't get stirred up."

Haylee and her friends seemed utterly sincere, totally confident in their convictions. If they remain so, though, they will be in the minority. According to Mark Regnerus, a sociologist at the University of Texas, nearly three-quarters of white evangelical teens disapprove of premarital sex, as opposed to half of mainline Protestants and a quarter of Jews. (Evangelical virgins, incidentally, are also the least likely to imagine that sex will feel good; Jews are most likely to cite pleasure as a reason to indulge.) Despite that, evangelicals are the *most* sexually active of those groups. They lose their virginity younger, at an average age of sixteen, and are less likely to protect against pregnancy or disease, perhaps due to a lack of education, or perhaps because preparing for intercourse would make their fall from grace appear premeditated.

Abstinence vows do have some impact, particularly among younger teens: according to sociologists Peter Bearman of Columbia University and Hannah Brückner of Yale, fifteen- and sixteen-year-old pledgers delay intercourse about eighteen months beyond their peers (though that's decidedly *not* "until marriage") and have fewer sexual partners. But the effect vanishes if more than 30 percent of those in a given community want in. Pledging has to feel special, like membership in an exclusive club. Hence, I suppose, the lure of abstinence swag: the rings, T-shirts, notebooks, wristbands, gimme caps, and other gewgaws that declare, "Don't Drink and Park," or "Keep Calm and Stay Pure," or simply, "True Love Waits."

So maybe it does, but not indefinitely and not for everything. Male pledgers are four times more likely to have anal sex than other young people, and pledgers of both sexes are six times more likely to engage in oral sex. What's more, by age eighteen, their resolve begins to crack; by their twenties, over 80 percent of pledgers either deny or have forgotten that they ever pledged at all. The only lesson that sticks is that they remain less likely to use contraception and drastically less likely to protect against disease. Having heard Pam Stenzel warn repeatedly that condoms are useless against infections and that taking birth control pills will leave a girl "sterile or dead," I guess I'm not surprised. Still, it's interesting that young adults retain the unsafe-sex messages of abstinence education even as they jettison the rest. The upshot is that pledgers have the same rates of STDs and pregnancy as the general population, even though they begin intercourse later and report fewer sexual partners overall. Nor is marriage fully protective: female pledgers married younger than other women, but even those who had never previously had intercourse (about 12 percent) tested positive for STDs at the same rates as married nonpledgers.

Folks such as Wilson and Stenzel like to say that waiting for your one true partner will make sex not only holier but hotter. The chemicals your brain releases during sex, they explain, will bond you to that one person, training you, Pavlov-style, to feel aroused and sensual whenever you are together. It's a romantic notion, but, again, it does not appear to be true. A 2014 study of young evangelical Christian men offered a more objective glimpse into the post-abstinent marriage bed. It turned out the men couldn't shake the idea that sex was "beastly" after the prohibition against it was lifted. They were surprised to find themselves still beset by temptation: pornography, masturbation, other women. What's more, back when they were single, they had the support of other abstinent men. Once wed, they found that talking to friends about sexual problems was considered a betrayal of one's wife, and they had no idea how to communicate with their spouses directly.

A young woman who had taken a virginity pledge in the Baptist Church at age ten told a similar story on the website xoJane. After marriage, she couldn't let go of the shame and guilt that had been drummed into her: "Sex felt dirty and wrong and sinful even though I was married and it was supposed to be okay now," she wrote. "Sometimes I cried myself to sleep because I wanted to like [sex], because it wasn't fair. I had done everything right. I took the pledge and stayed true to it. Where was the blessed marriage I was promised?" Meanwhile, a 2011 survey of more than 14,500 people revealed that those who had fallen away from religion were more sexually satisfied and felt less guilty about their sex lives than they had when they were believers.

AT THE BALL, the girls and their fathers rose from their tables, looked into one another's eyes, and exchanged vows. The girls committed themselves to purity. The men promised to "cover" their daughters, to lead, guide, and pray for them. The girls recited the following pledge: "Knowing that I'm worth waiting for, I make a commitment to God, myself, my family, my friends, my future husband, and my future children, to a lifetime of purity, including sexual purity, from this day, until the day I enter into a biblical marriage relationship." The couples then gathered at the back of the room. Each pair linked arms and, one by one, walked down the center of the dance floor, almost as if in a wedding. The fathers plucked silver tiaras from a basket and "crowned" their daughters; the girls then chose a white rose from a second basket.

Brittan introduced me to Dave, a divorced thirty-nine-year-old entrepreneur, who was there with his fourteen-year-old daughter. "As a dad, what I want for her is the best life she can possibly have," he told me. "And the truth is this: Whatever we do between the time we start becoming a young woman or man until we actually get married, whatever happens, whatever takes place in every relationship you have, whether it's physical, emotional, or mental—every experience you have you're going to bring that into a marriage. Purity can actually cut off at the root a lot of future pain. Instead of having to be healed of something, isn't it better not to get sick in the beginning? Who can argue against that?"

Dave should know, he continued. He faltered before his own marriage, something he regrets and blames, in part, for its ultimate failure. "I went off to college and was on my own," he said. "And I got off track. I did not surround myself with like-minded people. There was a lot of heartache and a lot of pain. That's why I think this is so flipping important. We get told all the time no one will be abstinent, there's no way they'll do it. Why? It's a choice." He pointed to his daughter, who was standing silently beside him, twirling her white rose. "If someone put a gun to her head every day and said if you lose your purity, I'll shoot you, I guarantee she wouldn't lose her purity. It's all about choice."

Dave did not, at least on the surface, hold a double standard. Abstinence, to him, was as important for males as females. He planned to serve as a role model to his children, remaining chaste until (or unless) he remarried. He expected "purity" from his sons as well. Again, his concern seemed less about sex than the pain wrought by emotional intimacy—pain that others may consider essential to personal growth, to developing mature ideas and expectations of relationships.

Listening to Dave, it occurred to me that the idea that purity would protect either him or his children from divorce—that practicing the skills of emotional or physical intimacy before marriage threatens rather than enhances a partnership—seemed as much a fairy tale as the fake crown he'd just placed on his daughter's head. I've been married nearly twenty-five years. Virginity, by the time of our wedding day long gone, was not something special or cherished my husband and I gave each other; our love and commitment were. That's true for all the long-term married couples I know; it was equally true of everyone I know who has divorced. What's more, if Dave really wanted his children to marry for keeps, he might want to start checking the real estate listings in liberal bastions such as New York, Boston, or San Francisco. Statistically, the strongest factor predicting higher divorce rates in any given county is its concentration of conservative or evangelical Protestants, in part because they marry and have children younger. Taboos against sexual experimentation and emotional intimacy may, then, boomerang on parents such as Dave, pushing their children to wed someone incompatible or before they're ready so they can have an openly physical relationship.

It's easy for those of us who think pledging is wacky to feel a little smug. Yet it occurred to me that these girls who were "virgins for God" weren't really so different from those who imagined virginity as a "gift" or even those who saw it as an embarrassment: they all believed that *one* sexual act would magically transform them—for better or for worse—and they all risked harm to their sexual and emotional development as a result. They all based their worth, calibrated their self-respect, and judged other girls' characters (tacitly or overtly) based on what was happening, or not happening, between their legs. And they all were still fundamentally defining themselves by their sexuality: by whether, when, where, with whom, and how many times they'd had intercourse.

By focusing on virginity, young people minimize (and often rush through) other forms of sexual expression, denying themselves the very opportunities for knowledge and experience that they seek. After all, moving slowly and intentionally with a partner is not only incredibly sensual, it's vital to learning, truly *learning*, about desire, pleasure, communication, mutuality, intimacy. That's ultimately far more life-altering than "achieving" intercourse. " 'Experience' is a stupid way of thinking about it," said Dennis Fortenberry, professor of pediatrics at the Indiana University School of Medicine and one of the foremost researchers on adolescent sexuality. "If you think of it as a *pool* of experiences of closeness, warmth, desire, attraction, arousal, touch, orgasm—all those are part of the possibilities of sexual learning. That's what young people should be doing. Learning about the incredibly nuanced thing we call sex that we assume will be part of their lives in different manifestations for the next sixty years or so. I don't think I'll see this in my lifetime, but what if we could even begin to think of actually saying to kids, 'Spend a year or two having oral-genital sex with people that you want to do that with and really get to know what that's about and then figure out what might follow.' "

I had walked into this ballroom unsettled, to say the least, by the white dresses, the wedding motif, the idea of fathers being made the guardians of girls' "sexual purity." The fathers were even given a Lucite-encased sixpence to keep as a symbol of their daughters' virtue, until the girls' wedding day (as in "something borrowed, something blue, and a silver sixpence in her shoe"). What could be more patriarchal, more regressive? At the same time, the sexualization so rampant in secular culture, which measures a woman's value first and foremost by how "hot" she is, is little better. I utterly, vehemently disagree with how they approach it, but like me, these parents only want what's best for their daughters; in their own way, they believe they're helping their girls combat modern pressures and degrading stereotypes. Brittan talked to me about the "pornography epidemic" and the importance of "empowering" young people to "navigate the assault of sexuality everywhere they go" so that they can make ethical, responsible, "healthy sexual choices." Like me, she believed that we should educate our children about sex "in a very direct way." It was all the same language, yet the intent was completely different. To me, purity and hypersexualization are flip sides of the same coin. I'd rather girls were taught that their sexual status, regardless of what it is, is not the measure of their personhood, their morality, their worth.

The dads and daughters, having completed the crowning ceremony and signed a "covenant" of purity, took to the floor for their "first dance," yet another ritual that mirrored a wedding. They looked so happy: the daughters basked in the attention of their fathers or mentors. I may not have agreed with the reason for the gathering, I may not have agreed with their message, but I did appreciate that fathers were communicating with their daughters at all, that they were taking time to deepen their bond with the girls: to create trust, to discuss ethics and values around sex. I interviewed more than seventy young women for this book: only two had ever had a substantive conversation about sex with their fathers. The rest just laughed when I raised the subject. Moms don't fare much better: even those who believe they've talked to their daughters about sex tend to overestimate the efficacy, openness, and comfort level of those discussions. Somehow, once parents stopped saying "don't," many didn't know what to say. So while it's easy to be appalled by the blatant sexism of Purity Ball dads—and yes, I absolutely was—I am equally appalled that the alternative to them seems to be total silence.

After a song or two, the dads drifted off the dance floor, while the girls kicked off their high heels. They jumped around in little scrums to "clean" pop songs such as Pharrell Williams's "Happy." As I slipped out the door, "Let It Go," the anthem from *Frozen*, came on. At the chorus, like young women everywhere, the girls flung their arms extravagantly wide and belted the words. The fathers looked on, smiling indulgently, apparently unaware that the point of the song—"No right, no wrong, no rules for me. I'm free!" and "That perfect girl is gone"—is that Elsa, the princess, is coming into her power, rejecting the restrictive, false morality that was imposed on her by her father, the king.

The Good-Person Checklist

Christina had known Brandon since kindergarten. They chased each other on the school playground, went to each other's birthday parties at the local skating rink. He won first prize in the middle school science fair, she took second. They shared their first kiss after the winter formal during their junior year. Over time, their physical intimacy deepened, but the specter of the Church was never far from her mind. "It was like, 'My boyfriend took off my shirt. What if other people find out?'" she recalled.

"Even now, I can logically talk myself out of those feelings, but it's all still there. There are degrees of shame and guilt that are probably permanently embedded in me. I wish that wasn't so. It haunts a lot of my actions." She paused thoughtfully. "But then, I don't know where the line is between how I was raised and what's just my personality. By nature, I'm a very cautious person."

Perhaps. Yet when I met Christina, she was planning a semester abroad in Botswana, which seemed pretty nervy to me. She'd also purposely chosen to attend a college that would challenge her long-held values, and sought housing that would push her even further. Christina's willingness to step so far out of the bubble of her upbringing—something that's hard for any young person to do regardless of her politics—struck me as admirable, even brave. She couldn't fully explain why she'd done it. It may have been because her parents weren't as conservative as her teachers. Christina's mother never contradicted the school's teaching on chastity, but she drew the line at its condemnation of homosexuality as a sin. "She just told me straight out, 'That's not true,'" Christina said. Beyond that, though, Christina always felt different from her peers. The other kids in her grade were white, and she resembled her Filipino father; she was the only Asian in the entire school. In middle school, boys teased her about the shape of her eyes, the color of her skin; it made her feel, even to this day, unattractive, undesirable. That sense of difference, of alienation, may have been enough to set her searching.

Christina expected her values to be tested when she entered college. "I knew I'd have to stick to my morals," she said. "If I didn't want to drink, I wasn't going to do that. If I didn't want to have sex with someone, I wasn't going to do that." Within a couple of months, though, she began to, as she called it, "loosen up": venturing out to parties, having a drink or two, making out with boys on the dance floor. "I guess I kind of did glamorize all that," she admitted. "I think I kind of envied the freedom of these girls who didn't have a lot of rules set up for them. I wanted to know what that felt like."

At one of those parties, early in the fall of her sophomore year, she met Ethan, a tall, gentle boy who, like her, was from a conservative community. They talked all evening, and found they enjoyed each other's company. At first she was hesitant to enter into a relationship, but within a month or so, they were dating exclusively, and by the end of October, they began having sex. "It was just very natural," Christina said. "I wanted to get to know him in that way, and he wanted to get to know me in that way. There was no pressure. It was totally my decision and all very partner-y."

Which is exactly how one would hope girls' experience of intercourse would be. Could that care and concern for a partner have been an unintended by-product of her conservative education? Was it simply because she was older than many girls at first intercourse? It's hard to say. Christina did credit her school with teaching an overarching ethic of kindness and respect for others—though apparently that didn't preclude people teasing her about her race. She also believed that since sex was off the table, boys in her class were forced, for the most part, to see girls as something other than sex objects. At the same time, that education left her especially insecure and ignorant about her body and its responses. "I didn't know *anything* before I got to college," she said. "I had no idea what a clitoris was. And there's still so much I don't know." Like what? I asked. "Well," she said slowly, "I worry about what's 'normal' in sex, but you can't ask because everyone is different. So I can't . . ." Christina trailed off. "I don't know. I don't know what's 'normal' for me. Like . . ." She hesitated again and then looked at me shyly. "Like is it normal never to have orgasms?"

Christina and Ethan were together for about six months. She never regretted losing her virginity with him, but once they broke up, she wondered, what now? "Am I going to be a person who only sleeps with people if I'm in a serious relationship? Do I want to make a rule that I'll go on a certain amount of dates with someone before I sleep with him? And if I do sleep with another person, that would bring my number to two. Do I care about that number?"

The "number" was a common source of concern among girls. Even those who felt that virginity was a vestige of another time wondered how many sexual partners was too many. (The "number," like virginity itself, included only intercourse—no one counted boys with whom, say, they'd had oral sex.) Losing their virginity in itself may not have tainted them, but was it possible to go too far? The stigma of the slut, the girl who was overly and overtly sexual, who allowed herself to be used, still held: their character could still be compromised, for themselves as well as others, by their sexual activity. "I guess I would feel icky if my number started to climb into the double digits," Brooke admitted. She glanced over at Christina, who was counting on her fingers, silently enumerating Brooke's lovers. "Stop that!" she snapped, laughing, and then grew serious. "I feel that sex is important. I don't want to have sex with people who don't mean something to me. And I'm not old enough yet to have had that many partners who do mean something."

Caitlin shook her head and pushed impatiently at her glasses. "I kind of don't feel that way," she said. "I feel like I could have sex with someone and it could mean nothing. I remember the first person I had sex with after the guy I'd been with for three years.

It was so surprising that it could feel . . . emotionally light, just fun and relaxed and easy.

"And what is that, anyway, to 'mean something'?" she continued. "Does it mean you have to love the person? Could it be about an out-of-body experience? Could it just be that this person was a good person and I appreciated how generous they were? Isn't that meaningful?"

Brooke shrugged, picking at her nail polish. "Maybe it's my own self-consciousness. For me, saying no is so hard under any circumstances, even to a favor for a friend. So I can see myself accidentally letting things escalate with someone I didn't want them to escalate with, and that wouldn't feel good to me. But I guess if I was turned on by someone who I wasn't into emotionally . . . I can't really imagine it, but that would be okay."

"It's such a relative thing," Christina mused. "Where I came from is so different than where you came from, so what sex means to me is so different. If a year ago I'd had sex with two people, I wouldn't have been okay with that. But now I am. So I think the 'meaningful' has to be a sliding definition both for each person and over time. And I think . . . I think I don't care anymore about someone's number. I mean, for safe sex, yes, but in terms of feeling like they're a morally better or worse person . . . I used to think the checklist of whether or not you were a good person was about 'are you drinking, are you smoking, are you having sex, are you loose in these ways'? That's not my checklist at all anymore. Because everyone has so much more depth and so many more dimensions than that.

"And I don't think I want to set lines for myself anymore, either," she added. "Because you'll be disappointed when you cross them. I have to trust myself to know what feels good and natural and what doesn't."

Caitlin was messing with Christina's computer and had cued up another Pam Stenzel video. This one was called "Definition of Sex." Stenzel was still pacing in front of the "High Cost of Free Love" sign, spieling like a Catskills *tummler*. She talked about a girl she'd met who'd had a "radical hysterectomy" at eighteen; her cervical cancer was diagnosed in ninth grade, caused by her contracting HPV in seventh. (While she warned, correctly, that condoms can't fully protect against HPV, Stenzel neglected to mention there is a vaccine, offered by pediatricians when children are eleven, that will. Nor did she mention that regular pap smears will effectively screen for abnormalities.) Then she began to talk once again about virginity. "I'm now going to give you the medical definition of 'sex,'" she said. (And right there a viewer should have been suspicious, since, as I've said, there actually isn't one.) "This is the medical line over which you can't step, and if you have ever stepped over this line, you have risked disease *and* you need to get tested, and don't you *DARE*! Don't you *DARE* tell anyone you're a virgin! Here is the line over which you can't step. Absolutely *no genital contact of any kind*. That's hand-to-genital, mouth-to-genital, genital-to-genital. Oral sex, which is mouth-to-genital, is sex. Hence the name 'oral sex.' And if you have had oral sex, you are not a virgin and don't you *dare* tell anyone you are."

The girls watching the video giggled and occasionally gasped in shock. Weirdly, though, I found myself agreeing with Stenzel, if not with her conclusions or her effort to shame and terrorize her audience. Our definition of "sex" is too narrow. I realize that it's idealistic to call for a dismantling of virginity for the sake of girls' health, but even questioning the implications of our assumptions about it has value. It is worth asking how putting this one act into a separate category is keeping girls (and boys) safer from disease, coercion, betrayal, assault; whether it gives them more control over their sexual experience; whether it encourages mutuality and caring; how it affects their perception of other kinds of sexual interactions; what it means for gay teens, who can have multiple sex partners without heterosexual intercourse. Again, this is not because that form of intercourse is no big deal, but because it's not the *only* big deal. I'd rather young people think of sex more horizontally, as Dennis Fortenberry suggested, as a way to explore intimacy and pleasure, than as this misguided vertical race to a goal. What if your first kiss were a form of virginity loss? The first time you had oral sex? What if it was first love? What if, as Jessica Valenti suggests in *The Purity Myth*, a girl didn't lose her virginity until she'd had her first orgasm with a partner?

Before leaving Christina and her friends, I asked how she would raise her own daughter if she had one. She pondered that for a moment. "There are huge holes in my sex education that I can't ignore," she finally said, "but at the risk of losing the other lessons that benefited me, I wouldn't wish to have done it differently. Still, I really want to have a more open discussion with my children. I can't quite imagine being at a level of saying, 'Okay, so this is what your clitoris is,' but then again, I'd want that for them if that would make them more comfortable in the world.

"I guess I would have to tell my daughter, 'It's totally your decision,'" she continued. "'It's whatever you feel comfortable with. But you have to be safe: there are these bad things that can happen in sex, but there are also benefits.' I would have to tell her, 'It's very much up to you and how you feel.' Because I think, in the end, it is the most personal of all decisions."

CHAPTER 4

Hookups and Hang-Ups

Holly, a sophomore at a private East Coast college, volunteered to talk to me for a specific reason. She wanted it known that some college girls, girls such as she, enjoyed the so-called hookup culture. "In books and articles they always say that if a girl sleeps around she'll get called a slut or that all girls only really want relationships," she said, sweeping her strawberry blond hair back over one shoulder. "Otherwise, it's just about how hookup culture is good for guys, and how they feel this sense of accomplishment when they've had sex with a number of girls. But I'll just put it out there: I feel accomplished after I have sex with someone that I wanted to have sex with. Last Thursday morning I woke up and apparently everyone in my sorority house knew I'd had sex because they'd heard the bed squeaking through the ceiling. And everyone goes, 'Holly! High five! You get it, girl!' I felt accomplished, just like a boy would. I felt like, 'I went out, I looked good, I showed myself off, and I got it last night. Good for me.'"

What's Sauce for the Gander

As with oral sex in the 1990s, discussions of the current "hookup culture" are fertile ground for good old-fashioned media-induced panics. The take-away from most reports tends to swing extreme: Hookups are terrible for girls! Hookups are liberating for girls! Girls are being victimized! Girls are going wild! Here is what they rarely say: young people are not, in fact, having more sex than they used to—at least, if you define sex by intercourse. The seismic tectonic shift in premarital sexual behavior really took place with the Baby Boom generation, according to Elizabeth Armstrong, a sociologist at the University of Michigan who, with her colleagues, has conducted the most comprehensive research on college student hookups. That was when the introduction of the Pill, the rise of the women's movement, and relaxed attitudes about supervision of "coeds" ignited the sexual revolution. Nor did today's young 'uns invent the concept of casual sex. What has changed, however, among college students and increasingly among high schoolers, is that when relationships do occur, instead of starting with a date, they often begin with noncommitted sexual contact. Rather than being a product of intimacy, then, sex has become its precursor, or sometimes its replacement. That's what is meant by the term *hookup culture.* "Casual sex was happening before in college," said Debby Herbenick at Indiana University's Kinsey Institute, "but there wasn't the sense that it's what you *should* be doing. It is now. I have students who say people should be able to have no emotions in sex, and if you can't, there's something wrong with you and you're missing out."

The word *hookup* itself, as I've said previously, is ambiguous, indicating anything from kissing to oral sex to intercourse to anal sex. To make things more confusing, there are different types of hookups: one-time hookups, repeated hookups, exclusive hookups, "friends with benefits." The only common thread is that there is no thread—or, more correctly, no *strings*: no emotional commitment, no promise by either partner of anything beyond the moment. According to the Online College Social Life Survey, which included some twenty thousand students at twenty-one universities, 72 percent of both male and female college students hook up at least once by senior year, with the average number of partners being seven. The behavior is most typical among affluent white heterosexuals and least common among African American women and Asian men. Twenty percent of college students hook up ten times or more by senior year; 40 percent hook up three times or fewer. Only a third of these hookups included intercourse; another third involved oral sex or some form of manual genital stimulation; the rest consisted of kissing and what my grandparents would have called "heavy petting." So it's not exactly the fall of Rome out there. Kids themselves tend to overestimate the sexual activity of their peers, again, perhaps driven by media "scripts"—from the 92 percent of songs on the *Billboard* charts that are about sex to movies such as *No Strings Attached* and *Friends with Benefits* to TV shows from *Pretty Little Liars* to *Vampire Diaries* to *Awkward* to *Grey's Anatomy* (Mindy Kaling, creator and star of *The Mindy Project*, has joked that her eponymous character has dated more men in a few seasons—making out with thirty of them—than she, the real Mindy, has in her entire life). There is also the rise of hookup apps such as Tinder, that portray millions of people as blithely bed hopping. But overstating the amount of sex going on is not young people's only perception-reality gap: when Herbenick anonymously polled the one hundred fifty students in the Human Sexuality class I had visited, over 70 percent of both sexes believed that their classmates solely sought hookups, while less than half believed others were interested in relationships. The truth is, nearly three-quarters of the boys and 80 percent of the girls said they'd prefer a date to a hookup, and nearly 80 percent of both sexes would like to be in a loving relationship within the next year.

Some girls, such as Holly, reported feeling affirmed by hookups, released from emotional responsibility for their partner, free to acknowledge straight-up lust. At the same time, the actual sex? *Meh*. Girls' physical satisfaction in hookups tends, once again, to be secondary, an afterthought. They are considerably less likely, for instance, to receive oral sex in casual encounters, and when they do, it's rarely to climax: only 17 percent of women reported orgasms in first hookups that included oral sex alone, as opposed to 60 percent whose most recent cunnilingus experience was in a relationship. (Men in hookups, incidentally, overestimate their partners' orgasms by a third to a half.) In hookups involving intercourse, 40 percent of women said they'd come (half the rate of men who did), as opposed to three-quarters in serious relationships. Orgasm may not be the only measure of sexual satisfaction—girls sometimes complained to me that the pressure from boyfriends to "achieve" climax stressed them out, especially when they were sexually inexperienced—but since young women are up to six times more likely to say they enjoyed an encounter (either in a relationship or a hookup) when they did come, neither is it irrelevant. Perhaps one could argue that it takes time for men to learn a female partner's body and responses, but it also requires interest—and basic respect. Young men routinely express far less of both for hookup partners than for girlfriends or even "friends with benefits." As one boy put it to Armstrong and her colleagues, "In a hookup, I don't give a shit." Women were equally invested in their partners' pleasure either way. That may partly explain why 82 percent of men said that the morning after a hookup, they were generally glad they'd done it, compared to 57 percent of women.

Even so, 57 percent is a lot of girls, enough to show pretty clearly that hookups neither are driven by nor benefit only boys. As the age of first marriage rose and the idea of finding one's husband during college became an anachronism, Armstrong and her colleagues found girls' willingness to devote time to relationships waned. With years of single life still ahead of them, many want to focus their energy on "self-development": pursuing academic, personal, and professional goals or hanging out with friends. Parents, too, have urged them to focus on ambition rather than romance. Hookups allow them to do all that while still enjoying an active sex life. Besides, how many times can you—or do you *want* to—fall in love? Hookup culture, then, acts as a kind of buffer, a placeholder until the time for more official adult partnerships begins. The girls I met often claimed to be too "busy" for relationships. On one hand, it was heartening to hear that their lives didn't revolve around men. Yet it was also hard to imagine a time when that "busyness" would abate—it would arguably become more intense after college, when they'd be career building or attending graduate school. What were they so busy doing, anyhow? It's not like they had to shop for food, prepare their own meals, or pick up their children at school. While I was all for broadening possibilities, the idea that romance and ambition were mutually exclusive troubled me. It sounded a bit too redolent of "you can't have it all," a phrase that blames individual women rather than structural inequities for our struggles at work and home.

"There's this idea now that identity is built independent of relationships, not within them," said Leslie Bell, the psychotherapist and author. "So only once you're 'complete' as an adult can you be in a relationship. It's an interesting shift from earlier academic thinking and folk wisdom—that women are naturally relationship-oriented and develop within them more than they do independent of them." Bell isn't opposed to hookups, but found that her own subjects, who were five or ten years older than mine, weren't having the experience of trying out love, intimacy, vulnerability, or self-advocacy with a partner. Their adulthood and independence were based on denying rather than expressing emotional connection through sexuality. "It's all about the importance of not getting played," she said. "Why isn't there much discussion about going through a bad love experience and learning from it? Why aren't there as many stories about the importance of taking risks even if you do end up feeling played? It's like a perversion of relatedness and interdependence—as though for women to participate in a relationship will always mean a loss of self."

Listening to Bell, I recalled a conversation I'd had with Mackenzie, a sophomore at a Bay Area high school dominated by hookup culture. She was going through a rocky patch when we met: her boyfriend of a year had just cheated on her, making out with another girl while drunk at a party, and she was conflicted over whether to break things off. She was often teary as we talked, describing ways she'd "lost herself" in their relationship. "I'm not saying that's all a negative thing, though," she added. "I've learned a lot about myself, too. I've learned that I have so much to me. I have a lot to give. Also I learned a lot about myself and vulnerability. I can love very deeply, and I think that's a good thing. I've learned a lot about my body, about my mind—just being with someone else, hearing their views on things, being intimate. I'm still learning. I'm learning what it's like to deal with heartbreak and someone you believed would never hurt you and he did. All of that."

On the college campuses I visited, hooking up was considered the ticket to a social life, to enjoyment, empowerment, even to a potential relationship. The girls who opted out, especially freshmen, could be left bored and lonely on a Saturday (or a Friday or a Tuesday) night. What fun was that? Their objections were usually not moral: they didn't think that girls who hooked up were "sloppy" or indiscriminate so much as that casual sex seemed emotionally hollow, potentially unsafe, and, sometimes, unhygienic. Becca, for instance, a freshman at an East Coast private school, had been nicknamed Grandma by her friends because she was often in bed by nine. She'd hooked up plenty of times when she was younger—making out with boys at the private Jewish middle school she attended, performing oral sex for the first time in ninth grade, losing her virginity at fifteen in a haze of weed and alcohol. Those experiences left her feeling lousy. Since early senior year, she'd had a steady boyfriend with whom she was in love; she remained committed to him even though he was at school in another state. "My friends have said, 'Bec, you shouldn't have a boyfriend when you're in college!' " she told me. "So, last night I went to a party and two separate people told me this sophomore guy wanted to fuck me. I was like, 'Great. He doesn't want to get to know me but he wants to *fuck* me?' I have found someone I genuinely love and I'm not going to let that go to hook up with random people. I mean, you want me to hook up with a bunch of guys and get mono? I don't understand." (Becca, it is worth mentioning, was the only girl I interviewed on her campus who was not sick with a nasty upper respiratory infection that students called the Sludge.) Similar to Sam, the high school girl who wouldn't spend time alone with her male friends, Becca also felt that the hookup culture was an obstacle to platonic relationships. "Like, I was hanging out at a frat house recently after a 'darty' [daytime party]," she said. "Just hanging out and talking to the guys, and one of the brothers was not shy in expressing his confusion over why I would do that, since I wasn't hooking up with anyone."

Sierra had her share of hookups in high school, too, but found them similarly unfulfilling. A freshman in college when we met, she'd been with her current boyfriend for nearly a year. "I used to think the sexual stuff was how you got to the emotional connection," she said. "But that's not true. The emotional connection comes first. That's what has made the sex so good. The first time we had sex, my subconscious was thinking, 'He's excited to do this not just for the sake of doing it, but to be doing it with *me*. To be doing it with someone he's going to end up loving.' He cares about how I am feeling. He texts me in the morning: 'Good morning! How are you today?' And if I text, 'I'm tired,' he texts back, 'Great. But how *are* you today? Mentally? Are you stressed? Are you happy? Are you sad?' It's knowing that we got to know each other, to know what makes us pissed off or happy or sad. It's that connection, that reassurance, that this isn't a 'hit and run.' We live in the moment and love every second, but it is absolutely the emotional connection before the sexual stuff that has made it worth it."

At the other extreme, or so I initially thought, was a freshman at a midwestern college who regaled me with tales of her sexual swashbuckling for nearly two hours, telling me how she rejected boys whose penis sizes "didn't meet my standards," or who were too heavy ("I don't like fat guys," she said). Yet at the end of our conversation, when I asked if there was anything she'd like to add, she hesitated, and almost in a whisper said, "philophobia."

I looked at her questioningly. "It's the fear of falling in love or being in love," she explained. "I read about it in a book. Sometimes I feel that's why I never get into an actual relationship. It's so hard for me to have an emotional attachment to people. I don't want to get hurt. So I just go from guy to guy, putting a barrier between me and others to keep that from ever happening."

I don't want to idealize relationships. While some girls had found love and joy within them, others had experienced manipulation and devastation. Becca had undergone two depressive episodes after splitting up with high school boyfriends. Mackenzie cried until she vomited when she discovered her boyfriend's recent betrayal, and had hardly eaten in days. Her schoolwork was suffering, too. More than half of physical and sexual abuse of teen girls by a romantic partner happens within a relationship, and those experiences prime girls to be victimized again in young adulthood. One girl I spoke with described how her tenth-grade boyfriend slapped her and flung her into a fence when she threatened to break up with him. Another girl, a sophomore in college, hadn't realized she could be—and was—raped by her recent boyfriend. Encouraging girls to explore sexuality within mutually caring, emotionally connected relationships is one thing; *insisting* on it is another. That can turn sex into a commodity that girls barter for the "safety" of commitment, and implicitly condone the shaming of those who don't comply.

There was no consistent attitude toward either hookups or relationships among the girls I met. They all, however, had to negotiate the culture of casual sex, whether they participated in it or not. They all had to find comfortable ground in a culture that was simultaneously fun and antagonistic, carefree yet riddled with risk. The question to me, then, became less about whether hookups were "good" or "bad" for girls than about how to ensure reciprocity, respect, and agency regardless of the context of a sexual encounter. That meant understanding the contours of girls' new freedom as well as the constraints, both physical and psychological, that remained.

The Happy Hookup

Holly, a Spanish and psychology major, revised her definition of "slut" for the first time when she was sixteen. She grew up in a mostly white, affluent, liberal East Coast suburb and attended a progressive, all-girls high school. Her mom told her to wait until marriage to have sex, but in Health class she learned about birth control and practiced putting a condom on a rubber model of a penis. (Again, though, the location of the clitoris, masturbation, and female orgasm went unmentioned.) In tenth grade, some of her friends began performing oral sex on their boyfriends; within a year or so, they were having intercourse. "My opinion had very much been, 'It's only those skanky public school girls who are doing that sort of thing,'" Holly said. "But if my friends were having sex, it had to be okay, right? So I had to reevaluate. I thought, 'That's fine; they've been dating for a year. They've built trusting relationships.'"

Holly, however, stayed both chaste and sober: a "good girl" who imagined she'd save sex for a loving relationship and alcohol until age twenty-one. When she did imagine having a boyfriend, her fantasies hewed to the romantic rather than the sexual—beaches and sunsets were usually involved. She entered college, she said, "very pure," but campus life quickly changed her. Her fourth night at school, she made out at a party with a guy she barely knew. It was fun. A week after that, she gave the same guy a hand job, and he fondled her breasts. "It was a *huge* thing for me," she recalled. "I touched a boy's penis! He touched my boobs! I was slightly overwhelmed. Because three weeks before, I would have said no. But I wanted to be doing this, although nothing more than this." By early October, she had happily hooked up with two more guys, making out on the dance floor and going back to their rooms. "I almost feel like I wanted the opportunity," she said. "Because in high school I never had the opportunity to hook up with boys. And in college I have this endless opportunity to do it, so I felt like I could."

Holly met Connor, who lived on her floor, at a school football game, and the two bonded over their politics—which were more liberal than those of many of their peers—and a mutual passion for *The Daily Show*. They began texting, and one night Connor asked if Holly and her friends would take him to a frat party. Freshman year was tough for boys on campuses dominated by Greek life. In order to "preserve the ratio" of girls to boys at a party—keeping the odds in the hosts' favor—frats limited the number of unaffiliated males allowed in. So unless a freshman guy was accompanied by a large enough group of women (three, four, sometimes more), he risked being turned away.

Holly showed me a picture of herself on a recent night out that she'd posted to Instagram. She was dressed in what I came to think of as the sorority uniform: a tight black miniskirt, bare legs, crop top, and stilettos. Her hair was flat-ironed straight, and she wore red lipstick and dark eyeliner. She looked like a different person from the scrubbed-face girl before me. "There are few times that I feel more confident about my body than when I wear a crop top and my boobs are showing and my legs are showing and I'm wearing super high heels," she told me. "I never feel more liberated than then. I'm proud of my body, and I like to show it off."

That phrase, "proud of my body," continued to bedevil me. On one hand, I admired the young women's bravado, their willingness to be overtly on the prowl, their refusal to be shamed for how they did or didn't dress. At the same time, only certain bodies were allowed to be a source of "pride," to be seen as sexual, to deflect shame, and Holly's had not always been one of them. As a freshman, she was twenty-five pounds heavier than when we first met—she'd dieted and worked out all summer to lose the weight—and her wardrobe had been considerably more conservative. "I would never have worn anything skimpy because I wasn't happy with how I looked," she said. "Presenting myself in skimpy attire would have had a very negative impact on my mental state, because there would be those people, especially boys, who would say, 'She's fat and she should wear something else.'" It's understandable that Holly would feel good about showing off the "right" body—it's affirming to attract male approval and even female envy—but it's hard to see her outfits as "liberating" when the threat of ridicule always lurks. One of her sorority sisters, for instance, had recently gained weight. "It's not that she *couldn't* wear skimpy clothes," Holly said. "But she knows how she would feel if there were asshole-y boys who were like, 'She's a fat girl.'"

On most of the campuses I visited, Greek life (or houses where athletes lived) was the center of the hookup scene. The twenty-six sororities in the National Panhellenic Conference are voluntarily dry. So it is the frats that host, control entry to, and provide alcohol for most parties. Fraternity pledges typically chauffeur groups of girls from freshman dorms or sorority houses to events (though not necessarily home again) that can offer endless variations on a single concept: young women as prostitutes. Themes include "CEOs and business hos," "workout bros and yoga hos," "lifeguard bros and surfer hos," "GI Joes and army hos." Girls who liked to party shrugged off those slights (similar to the way they ignored degrading lyrics in a favorite song) as a form of "boys will be boys," unconnected to how most guys acted "in person." Frats got in trouble only when their sexism became even more egregious or was mixed with racism: the Phi Sigma Kappa chapter at California Polytechnic was investigated in 2013 by the school's administration for its "Colonial Bros and Nava-Hos" party. (No violations of university policies were found.) The Sigma Chi chapter at Harvard raised hackles with a similar bash, called "Conquistabros and Navajos." Meanwhile, the Duke chapter of Kappa Sigma was suspended in 2013 after news broke of its racist "Asia Prime" party, whose invitation began, "Herro Nice Duke Peopre!!" (Duke frats have made headlines repeatedly in the past few years for such antics as inviting "all potential slam pieces" to a "Plan-B Pregame" party and sending an e-mail to female classmates requesting they arrive at a Halloween party dressed "like a slutty nurse, a slutty doctor, a slutty school girl, or just total sluts.") The Delta Kappa Epsilon chapter at Yale was banned from campus in 2010 after brothers gathered near the freshmen dorms and chanted, "No means yes, yes means anal!" Wtf and "My name is Jack, I'm a necrophiliac, I fuck dead women and fill them with my semen." Students protested in 2012 after the same frat's Amherst chapter had a T-shirt printed up for its annual pig-roasting party depicting a woman clad in a bra and thong tied up and roasting on a spit, an apple jammed in her mouth, her sides bruised, and a pig standing beside her. Its caption read, "Roasting Fat Ones Since 1847." In 2014 the Texas Tech chapter of Phi Delta Theta had its charter revoked for displaying a banner that read, "No Means Yes, Yes Means Anal!" at a party, along with a "vagina sprinkler" that shot water at guests.

WTF

The members of all those houses, as in most of the Greek system, were primarily white and affluent; somehow they believed that racism and misogyny marked them as rebels rather than merely the latest recruits to an entrenched old guard.

Young women are tacitly expected to repay their hosts' generosity with sex, or at least the promise of its possibility. "Every girl knows that when you walk into a fraternity house, your most valuable asset is your sex appeal," a junior at a private East Coast college told me. "Everyone knows you have to imply you'll have sex with guys to get them to give you alcohol, drugs, rides, whatever. Everyone plays this game—and since at my school we're all overachievers, we do it really well!"

Girls who pledge sororities at Holly's school were required to attend frat parties at least four nights a week. (There were "ragers" every night but Monday.) Before the main event, they would "pregame" with a different frat, socializing and drinking for an hour or two. Holly would typically have three or four beers at those occasions and sometimes also a couple of shots. The girls would then be picked up by a second round of pledge rides and driven to the real party. "In some houses, basically you get there, go down to the basement, grind with a guy, and go back with him. Just that fast. But at my favorite house, I talk to my friends, we play drinking games, we dance a little, we go back and smoke a little. Sometimes I'll just dance with my sisters, and that's a good time. And grinding is fun, too. It's fun to have a guy holding on to you like that. You don't have to hook up—and anyway, there's more girls than guys at parties, so not everyone can. But it's often a big hookup scene."

When I added it up, Holly was regularly downing three to six (or more) drinks in an evening. For women, four qualifies as binge drinking. She didn't consider herself a heavy drinker, and likely her friends wouldn't, either. Alcohol is endemic to hookup culture. Hookups aren't just lubricated by drinking; they are *dependent* on it, in order to create what Lisa Wade, an associate professor of sociology at Occidental College, calls "compulsory carelessness." As a sophomore at an East Coast university told me, "It's like the girls I know live dual lives. From Sunday night to Thursday afternoon we're in the library all the time, working really hard. Then comes the weekend. We all rip back shots in our dorms before a frat's *pre*-party. Like four to eight shots in about a half hour. That's pretty normal. And then it's normal to wake up next to some guy and not remember how you got there."

Alcohol, according to Wade, is how students signal to one another that the sex they're having is meaningless. For her own research, she asked eighty-four freshmen to submit weekly journal entries over the course of a semester about sex and dating on campus. "They talked about having sex while sober in these reverent tones," she said, "like it was an amazing unicorn: it was 'meaningful' in a way that drunk sex is not." Drunkenness had replaced mutual attraction as the fuel for sexual interactions in college: "In a morning-after recap," Wade continued, "it is a reason in itself to have had sex."

As with intercourse, the proportion of young people who drink has actually dropped over the past decade, but the *amount* that girls in particular (and white girls specifically) drink on each occasion has not. A 2013 survey by the Centers for Disease Control and Prevention found that one out of four college women and one out of five high school girls had binged within the previous thirty days; they typically binged three times a month, downing an average of six drinks on each occasion. Other surveys have found that nearly two-thirds of college women and over 80 percent of men had episodes of binge drinking, and linked the practice with disordered eating—sometimes called "drunkorexia"—among girls who try to restrict food intake to reserve their calories for alcohol. Eighty-nine percent of college students get drunk before a random hookup, averaging four or more drinks each time. Three-quarters get drunk before hooking up with an acquaintance. They're most likely to be the most drunk when the encounter includes some form of penetration: oral, vaginal, or anal; they're also most likely to express regret after such experiences.

The girls I talked to often spoke of "going crazy" as an integral part of "the college experience"; they sounded like they were all quoting from the same travel brochure. I'm not sure when that phrase began to refer specifically to drunken partying. Although I recall a certain amount of alcohol and weed when I was at school, if someone had asked me to describe it, I would have said the "college experience" was more about redefining myself away from my family through intense late-night talks with friends, exposure to alternative music and film, finding my passions, falling in love. But according to a blistering exposé by Caitlin Flanagan in *The Atlantic*, as tuitions have skyrocketed, universities apparently need to convince "consumers" (their prospective students) that it's worth the staggering debt they'll take on to attend. What better enticement than to position higher education as not only edifying, but off-the-chain fun? "Every moment of the experience is sweetened," Flanagan wrote, "by the general understanding that with each kegger and rager . . . they are actively engaged in the most significant act of self-improvement available to an American young person: college!" That's a far cry from the original purpose of universities: to train young men for the ministry, a process that involved asceticism, temperance, and chastity.

When I asked why they didn't hook up sober, girls would laugh and say that would be *awkward*—their catchall word (along with *uncomfortable* and, sometimes, *weird*) for any unpleasant emotion. In this case, what seemed to unnerve them was not only having nothing on which to "blame" their behavior, but the idea of being fully emotionally, psychologically, and physically present in a sexual encounter. "Being sober makes it seem like you want to be in a relationship," one freshman told me. "It's really uncomfortable."

That first night Connor tagged along with Holly, they both got tipsy and kissed on the dance floor. The next day, they attended a football game together. Within a week, she had given him oral sex, something she'd never done before. "It was like, 'Whoa! Where did this come from?'" she said. "He didn't even ask. I was slightly alcohol-induced, and I was like, 'Okay, I'm just going to go for it.' And I thought, 'You know, this isn't too bad. Why was I making a big deal about it?'" She paused, considering. "That was the moment, I think, when I became a lot less uptight."

Looking back on it, Holly believed she was "too generous" with Connor—she wanted to make him "happy," but he didn't seem to return the sentiment. "There was one night I asked, 'Do you want to give me oral?'" she said. "He went down on me for about a half second. Then he said, 'I just can't do this. It grosses me out.'"

"I mean, I had a good time," she continued, "but it wouldn't be about *me*. My orgasm was never a given. It was not as important. It was not part of the deal."

Two weeks after their first hookup, Connor asked Holly to be his girlfriend. She was thrilled. He never pressured her to have intercourse, she said; he told her to just tell him when she was ready. A month later, she was. She thought it would be "like the movies: this magical and beautiful moment." She even decorated her room with Christmas lights for the occasion. Instead, it hurt. A lot. "I made him stop. We kissed for a little while and cuddled and were cute with each other. And then I said we could try again. It lasted a little longer, but it still hurt too much."

Intercourse may have been a disappointment for Holly, but it still felt like an accomplishment, a milestone. After Connor left, she strutted into a friend's room blasting the song "I Just Had Sex" on her iPod (a somewhat ironic choice, given that the lyrics—"I just had sex, / And it felt so good, / A woman let me put my penis inside of her"—describe a guy who is comically oblivious to his partner). "I was like, I feel *so cool!*" she said. "I feel like such a grown-up! And I had shared this special moment with a guy who I liked and trusted and who I had feelings for and who had feelings for me. Also, I was sober—that was very important to me. I was not going to have sex the first time drunk. I wanted to be able to *experience* it."

Connor broke up with her two days later.

This was a boy who had compared their relationship to his parents' (who had also begun dating each other the second month of freshman year). He had talked about how much he'd miss her over winter break, which was still over a month away. He had asked her to be his *girlfriend.* Holly was devastated. She left school two days early for the Thanksgiving break, needing to get away.

When her parents picked her up at the train station, her mom looked her up and down. "You lost your virginity," she said.

"I asked her how she knew," Holly told me. "And she said, 'Look at you. You're a mess! I hope that's a good lesson for you about not giving your body away to just anyone.'"

Girls' ideas and attitudes about sex are shaped by family, media, friends, and their own experience. Holly had followed the contemporary rules of female sexual respectability, done everything she believed was "right," and she was betrayed. She responded by giving up on love and commitment. She wanted to be "not *feeling-less*, exactly, but not in a relationship." Besides, she was busy: doing her schoolwork, pledging her sorority, going to parties. She still planned to reserve intercourse for a committed partner, whenever that might happen. "I felt like"—she stopped and corrected herself. "I *still* feel like it means something, that you're intimately connected and really like this person and you're showing affection."

Since she didn't have a boyfriend, Holly invited a male buddy from her dorm to her sorority's winter date party in February. They arrived already loaded—she'd had six shots at pregames. After the party, she went back to his room thinking they would make out, but she was still awfully drunk. So when he said she was beautiful and that he'd like to have sex with her, she thought, "Why not?"

A few minutes later, she felt as if she'd snapped out of a trance. "I thought, 'Holy shit! I'm having sex and I'm not supposed to be doing this unless I'm in a relationship.'" Holly panicked, telling the boy she needed to stop. He urged her to stay, but she jumped out of bed and threw on her dress. Still barefoot, holding her shoes, she flung open the door of his room to find a group of young men standing directly outside, listening in. She ran to a friend's room and cried.

"I was so upset with myself that I'd had sex outside of a relationship," Holly said. "Which I eventually got over. Now I don't care so much about that. I just care that I know the guy. But back then, in my head, I was a skank. I was one of those skanks who just has sex with people. I was a bad person."

Everyone's Slutty Friend

A picture of a kitten hung on one wall of Megan Massoud's room. Above her pillow was a poster from *Pulp Fiction*, the one in which Uma Thurman lies stomach-down on a bed, her stiletto-shod feet crossed at the ankles, a cigarette dangling from the fingers of one hand, a pistol flung casually near the other. Megan's desk was littered with half-drunk bottles of Coke Zero, open boxes of cookies, and several shot glasses. I picked my way through piles of clothes heaped on the floor, cleared a chair of some laundry, and sat down, resting my feet on a polka-dotted hassock.

Megan, a sophomore at a midwestern public university majoring in economics, was tiny (barely five feet tall), with enormous dark eyes, a quick smile, and flat-ironed dark hair that she would absentmindedly braid and unbraid as we spoke. Her mom, she said, was a "generic white woman." Her dad, who was Lebanese, gave her a pink lipstick-shaped canister of pepper spray just before her freshman year; Megan kept it out as a joke. "He thinks I'm a virgin," she said, laughing.

Megan pulled on a cropped orange tank top and a thigh-grazing skirt that hugged her butt and fit tight across the stomach. She examined herself in the mirror from the front, from the side, from the back. "Does my stomach look big in this?" she asked a friend, who was standing in the doorway. "Don't fuck with me."

"I'm not fucking with you," the friend said. "You look hot. Like the skinniest fucking bitch." Megan looked at herself again, dissatisfied. "I never think about what I eat until I get dressed for a party. Then I think I shouldn't have had that extra doughnut," she said.

As she continued to dress, Megan told me about the Gender Studies class she was taking this semester. "I had never noticed that guy models in ads are always doing something—playing a guitar or driving—and girl models are just . . ." She struck a classic pose: head tilted, chin down, hand on hip, a coy smile.

I laughed. "You do that really well," I said.

"I haven't taken a picture without the head tilt since I was six," she replied. "I don't know where I learned it."

She looked again at her stomach, again at her butt. She changed her shirt. She pulled off her skirt, tried on a different one, decided the first made her stomach look better, changed back. "In my gender class I'm all, 'That damned patriarchy,'" she said. "But at night it all goes to shit. The only thing I care about is: 'Does this skirt make my ass look good?'" She grabbed her cosmetics pouch and headed to the bathroom. Although she hates makeup, she said, it's part of attracting guys' attention, so she swiped on dark lipstick and some smoky, sparkly eye shadow. She smoothed her hair with two hands (a brush in one, a comb in the other), put on a pair of four-inch heels, and doused herself with perfume. "It makes me feel less self-conscious to wear the outfit and the heels," she said. "I feel kind of like I'm swaggering, like, 'Yes, I am the baddest bitch in this room.'" The evening was cold, but Megan didn't take a jacket. Nor did she carry a purse. She held her keys and school ID in one hand (later she would lose both) and tucked her phone and iPod into the waistband of her skirt, which was tight enough to hold them fast. She looked in the mirror one last time, turning to check her butt, and tugged down the hem of her skirt, a gesture she'd repeat every few minutes throughout the evening. She grabbed a bottle of vodka to share at pregames, and headed out the door. It was nearly ten o'clock. Her goal? "To get really drunk and make out with someone," she said cheerfully. "Because what's the point of a night if you aren't getting attention from guys?"

The stigma of "slut" didn't disappear with the rise of hookup culture. Its criteria just became ever more elusive. Girls routinely told me they hated the word, that they never used it, didn't "slut-shame" their classmates (though in truth, they often did). At the same time, they policed themselves. Some, like Holly, would continually revise, rather than discard, the definition of "skank" as their own behavior changed. Others, such as Megan, took "slut" on as a badge of honor, or at least tried to. "I'm the slutty friend," she told me gleefully when we'd first met. "I find it liberating. I love being the crazy one. If someone is going to judge me for what I do, then fine, judge me. I don't care. Fuck you if you think you're better than me just because you don't have sex that much. I feel bad for you if you don't have sex that much because sex is awesome. I'm not saying that every time I go out, I hook up with someone. That's definitely not the case. But it's more fun to not control myself. Not to worry how it will look. And in college, nobody gives a shit."

Like Holly, Megan described her behavior as "liberating," even as she struggled with its limits. During another conversation, she insisted, "I'm not a slut. Some people probably would consider me one, but I don't consider myself one because I don't carry myself like that. . . . When I think of a slut, I picture that girl who has the really thick black mascara and smoky eyes and wears two bras to push up her boobs." On another occasion, she told me, "I love being single," and a few minutes later confided, "No boys want to date the slut." Back and forth she went, between resisting and submitting to age-old ideas about girls' sexuality. Talking to Megan sometimes felt like watching someone trying to shore up a sand castle whose walls kept collapsing. Megan had less transcended limits than tried to legitimize herself within them, despite them. "I think," she told me at one point, "that every girl's goal is to be just slutty enough, where you're not a prude but you're not a whore. Yeah, you have your one-night stands. Yeah, you're experienced. But you're not sleeping with every guy in the fraternity. You're not making brothers 'Eskimo brothers'"—when two or more fraternity members have inter-course with the same girl. "Finding that balance is every college girl's dream, you know what I mean?"

Like Holly, Megan had her own agenda for our conversations: she, too, wanted an opportunity to make sense of a sexual history that had progressed quite differently from how she'd once ex-pected. Also like Holly, she described herself as a "good girl" in high school—not even kissing a boy until she was seventeen, by which time she was eager to move forward. "I really wanted to get rid of my 'firsts' with a boyfriend," she said. "And all my friends had already kissed guys, already given blow jobs. I was behind." During four months of dating her first boyfriend, she "caught up," performing oral sex that was never reciprocated. "I didn't even think that was an option," she said. She lost her virginity the summer before college with another guy she was dating, though, she said, they were never "Facebook official." She was relieved to get first intercourse over with, and remembers the experience fondly.

Megan had masturbated since she was a young teen. She had no difficulty reaching orgasm on her own, but had never cli-maxed with a partner. "A lot of guys don't do enough foreplay," she explained. "They just get to sex really quickly. And then, af-ter a while I get tired, and I know they're doing their best, so I just fake an orgasm to end it, and then I'm like 'Oh, that was *so* good.'" Most of the girls I talked to had faked an orgasm now and again; that seemed unfortunate, though not unusual. But according to *The Sex Lives of College Students*, the number who fake has been rising steadily, from less than half in 1990 to 70 percent today. That may at least in part explain the gulf between the proportion of boys who think their partner has come during an encounter and the percentage of girls who actually did. Girls feigned climax because they were bored, they were tired, they were in pain, they wanted the night to end. They were often, like Megan, protecting their partners' egos, or felt pressured to be perceived as enjoying sex even if they weren't— especially since pleasure was presumably the whole *point* of a hookup. They also faked because they didn't, or couldn't, ask for what they wanted in bed. A few were starting to question whether the practice was counterproductive. "I haven't really cared enough about the people I've been with to invest the time in training them in how my body works and what I like and don't like," a sophomore at an Ivy League college told me. "But now I'm going to put in the effort. Because I feel like I owe it to other girls to do them the favor of bringing these things to guys' consciousness. And why am I using my time like this if I'm not even going to enjoy it?"

Megan, like Holly, had her first college hookup within days of arriving on campus. The sex, she said, was "pretty terrible. He was the thrusting type, you know, jack-hammering me until I faked an orgasm, and then he went to sleep." Even so, she said, she continued hooking up with him semiregularly over the next two months. I asked her why she went back when the sex was so bad. She shrugged. "Sex is always good on one level," she said. "And whenever I get drunk, I hate going home alone. It's like, I need a boy or a burrito, you know?"

When we'd first met, midway through her sophomore year, Megan pulled out her period tracker app, where she had logged hookups that included intercourse. She'd had twelve partners, she said—though, if anyone asked, she reduced it to a more socially acceptable five. She preferred to remain "blissfully ignorant" of how many hookups had included only oral sex. "Giving a guy a blow job is something I don't really consider a big deal," she said. "Like, this one guy, when I go to his frat he'll say, 'Hey, Megan, do you want to come see my room?' And I'll give him a blow job and we'll make out. I told him, 'I like this casual thing we have going.' He's like, 'I know, me too.' I don't even have his phone number."

It was clear to me what he was getting out of that arrangement, I told Megan, but what was she getting? She shrugged. "I guess I could ask that every time I have sex. 'What am I getting out of it?' Guys tell me I'm really good at blow jobs, probably because I have a lot of practice. I really like kissing him. It's exciting, it's an adrenaline rush. And it's like, at least I'll have company. At least he'll appreciate me, even if it's for that fifteen minutes. I'll have someone to hang out with, and make out with, and make me feel special."

When the Fun Stops

Holly needed a guy. That's what one of her sorority sisters thought. So she asked her boyfriend to introduce Holly to his frat brother Robert. The four of them would go out to lunch, they'd go on double dates. Holly thought Robert was sweet, but she wasn't especially interested in him, either romantically or sexually. Still, simply by virtue of being thrown together so much, they got to know each other, and one night, at a party at his frat, they began making out on the dance floor. A little while later, she "found herself" in his room, doing "everything but intercourse." She had a wonderful time. "Oral sex both ways," she said, "which was a big deal for me." Robert walked her back to her dorm afterward. Even though she was hammered, she said he was a "gentleman and didn't take advantage of that and have sex with me."

The school year was winding down by then. She and Robert texted each other through finals, went for a couple of walks, made out. She had no interest in anything more; she was just enjoying his company. One night, after midnight, they snuck into an academic building and hooked up in a classroom. She'd had two beers, but said she wasn't particularly drunk. Neither was he. Again, they did "everything but intercourse," though this time it was mainly because he didn't have a condom. "Weirdly enough, I really wanted to have sex with him," Holly said—perhaps because he was the first guy who seemed authentically invested in her physical pleasure. "It was good that we didn't, though," she continued, "because I would have hated myself. I would've thought, 'Look, you've only started to get to know this guy. You need to know him better.'"

Over the summer, Holly tried to talk to her mom about birth control. She wanted to go on the Pill. "I told her it was safer in the social environment that I was in to have it, in case something happened. But she said, 'Well, you shouldn't want to be having sex. You're not in a relationship. You're nineteen years old.' And in my head I was thinking the opposite: 'I'm nineteen years old, I'm not in a relationship, and I *want* to be having sex!' She has no idea. If I told her what I've told you, she wouldn't let me come back to college. She'd say I was 'one of *those* girls.'"

Something else happened that summer, too. Holly had never before masturbated—it wasn't something she thought girls did. A few of her sorority sisters, as a joke, had given her a vibrator for her last birthday. One day, home alone and bored, she decided to give it a try, and she had her first orgasm. She spent the rest of the summer exploring her body. "It was cool!" she said. "I was able to learn all about myself without having to feel the awkwardness of trying to direct someone else." Girls often told me their first orgasm was transformative, whether they experienced it alone or with a partner. Why wouldn't it be, given their dearth of education on the subject? "The first time I had an orgasm I cried," one high school senior told me. "I *cried*! It was so powerful. I think it really helped me grow as a person."

Holly started her sophomore year with a new sexual standard. Still uninterested in a "serious" relationship but eager to experiment, she decided she would have intercourse only with someone she knew in a situation in which she felt safe. "Like, not in some weird room somewhere where you can't get help if you needed it," she said. Also, condoms were nonnegotiable. Then, one night, she did three shots at pregames and another three at a party. Then she had a "Jäger bomb," a shot of Jägermeister dropped into a beer. She followed that up with a Red Bull. Mixing energy drinks with alcohol leaves a person feeling deceptively sober—or "wide-awake drunk": college-age bar patrons who mix caffeine and alcohol, for example, leave drunker than their peers yet are four times more likely to believe they can drive. Maybe that's why Holly's sorority sisters, who are supposed to "look out" for one another, thought she seemed fine. Or maybe they were in no state themselves to notice. Either way, that drink was the last thing Holly remembered that night.

MEGAN WAS PLAYING beer pong at a low-key party just after winter break when a sophomore named Tyler began flirting with her. When her friends got ready to leave, around two in the morning, he asked Megan to stay.

"I'm not going to have sex with you," she told him.

"That's cool," he replied. "We'll just kiss and cuddle."

Megan's friend caught her eye one last time, silently double-checking her decision. Megan nodded. She wasn't too drunk, and she was having fun with Tyler.

They held hands and chatted as they walked back to his frat, getting to know each other. He seemed sweet. As soon as they got inside, though, his manner changed. He rushed her upstairs to his room and into his bed. They made out, and she started going down on him, but he kept pushing for intercourse. Megan said no. He pushed harder. Megan claimed not to have birth control, thinking that was a good, inoffensive excuse, one that wouldn't hurt his feelings. Instead, he grabbed a condom, held her down, and entered her. "I just kind of laid there," she said. "I thought maybe if I'm really shitty at sex, he'll just stop. At one point he asked if I wanted to take a shower together, and I was like, 'Well, we already had sex. What's the point in saying no now?' I just kept trying to make it better, to psych myself into thinking it wasn't what it was."

In the shower, Tyler kissed her roughly, then pushed her up against the tiles and began having sex with her from behind. She turned up the hot water tap all the way, hoping that would make him stop. It didn't. He switched to anal sex. "I told him he was hurting me, and he was like 'Oh, I'm sorry,' but then he'd keep going. His frat brothers actually came into the shower and saw us and laughed." She asked Tyler to stop twice more; finally he did. Not knowing what else to do, she spent the night. The next morning, when he dropped her off at her dorm, she told him, "Thanks, I had fun." She still doesn't know why she said that. A friend stopped by her room to find out how her night had gone.

"I think I was raped," Megan said.

THE CAMPUS PARTY scene can be exhilarating—if it weren't, no one would participate. But as Armstrong and her colleagues have pointed out, it also facilitates rape. Women, not men, wear body-baring outfits. Women, not men, relinquish turf and transportation. Women, as females and often as younger students, are expected to be "nice" and deferential to their male hosts. A "fun girl" wouldn't make a scene just because a guy grabbed her ass or held her down and grinded against her; she'd just find a way artfully and politely to disengage. "Fun girls" also drink freely— alcohol gives them license to be sexual, loosening inhibitions while anesthetizing against intimacy, embarrassment, or accountability. It can also undermine their ability to resist, remember, or feel entitled to report sexual assault. The manipulation of the party culture is both systematic and invisible, Armstrong writes, seemingly part of the continuum (if at the extreme edge) of acceptable "crazy" collegiate behavior. Since victims have a hard time convincing anyone, including themselves, that a crime has actually occurred, it is also generally consequence-free.

Holly woke the next morning with no idea where she was. There was a guy next to her in the bed, a senior she knew only by name and didn't remember seeing at the party. There was also a used condom on the floor.

"Do you remember what happened last night?" he asked.

She shook her head.

"We had sex," he said.

The boy lived several blocks off-campus, and claimed his car had broken down. So Holly, still dressed in the party clothes and high heels that had made her feel "proud of her body" the night before, made her way back to her sorority house alone. The so-called walk of shame is another aspect of hookup culture that calls out only young women's behavior, since boys often wear the same clothing at parties that they'd wear during the day. Sometimes girls borrow something from a sexual partner (though they may never have occasion to return it), but as Megan told me, "Everyone knows when you're in 'shacker clothes' and they'll heckle you when you cross campus, like, 'Ohhh! How was your night last night?'" Again, such harassment is typically leveled only at girls.

Holly spent the rest of the day in sweat pants, crying and watching TV while her roommate hugged her. That was just two weeks before we met. "I'm not going to let it ruin my life," she told me, her voice stalwart. "It's not something that defines me. It was just something that happened, and I can't get that drunk again."

While getting blackout drunk is never a good idea, and it seemed only natural for Holly to want to regain some sense of control, it troubled me that she placed all the blame on herself, on her drinking, rather than on the boy who took advantage of it. "I'd like to say he didn't know how drunk I was," she said. "But I don't know. My friend who is in an organization that fights rape on campus said that by definition I couldn't consent, so I was raped. And I almost . . ." she paused. "Not that I wish rape upon myself, but I hope I wasn't sitting there saying, 'Yeah, I want to have sex!' Because that would go against everything that I've said about not having sex with a random person." She shook her head and sighed. "I guess I'm fortunate that I don't remember."

I had no way of knowing, when I met them, that Megan was a rape victim or that Holly may have been. I didn't ask about nonconsensual sex in my recruitment e-mails, and it wasn't, they each said, what had motivated them to talk to me. A report by the Justice Department released at the end of 2014 found that, despite the growing national awareness of campus sexual assault, only an estimated 20 percent of college victims report the crime, a markedly lower rate than nonstudents the same age. They're inhibited by fear of reprisal, shame, self-blame, or the belief that reporting would only make things worse, especially given the historically low rate of campus assailants who are punished. Also by the deliberate muddying of consent that happens at parties. Mariah, a junior at a private university in the South, urged me not to demonize the Greek system. "I'm an intelligent woman," she said in an e-mail. "If all a sorority did for me was make me vulnerable to sexual assault and alcohol poisoning, I'd have bailed by now." She had made the dearest friends of her life among her sorority sisters, girls she described as "involved," "inspiring," and "brilliant." Yes, she said, the Greek system was "heteronormative" and riddled with racial and gender inequalities that needed to be addressed. "But I firmly believe," she wrote, "that sororities are, and can be, a wonderful experience, a vehicle for change, and a bastion of feminism on modern college campuses."

At the same time, though, she felt that she and her sisters were being "crushed" by a campus hookup culture in which drunken frat boys felt free to touch, kiss, or rub up against them without permission. ("You're supposed to swat them away like flies," she said.) Girls could quickly slip from feeling emboldened by sexiness to feeling objectified: like things to be used and consumed. Boys, too, could feel confused, uncertain: eager to fit in, yet struggling with assumptions about masculinity, sex, coercion, conquest. They could misinterpret mixed messages, or be too drunk themselves to realize a partner's state—both may wake up the morning after unsure of who they're with or what went on. "No one here knows what rape is," Mariah wrote, neither the boys nor the girls. "Would I know if I was raped? Maybe if it was a stranger in a dark alley, yeah, but otherwise, I'm not so sure."

I was surprised, then, to hear that Megan, at the urging of a campus therapist, had pressed charges against Tyler through her school's office of student ethics. The investigation took the entire second semester. Megan told her story repeatedly. Her friends gave statements about how much she'd changed since that night, growing depressed, unable to concentrate, how she dropped a class and was drinking more than usual. Tyler gave his version of events as well. When asked when, precisely, he believed Megan had given consent for intercourse, she recalled him saying, 'Well, she gave me a blow job. I pretty much call that consent.'" That had infuriated her. "I was giving him the blow job to end it, not to start something. I told him I did not want to have sex. I told him I did not have birth control. And he just hopped out of bed, put on a condom, and raped me."

What she suspects ultimately made her case was not so much what either she or Tyler had said, but that Tyler's own frat brothers turned on him, admitting that he could be aggressive, even violent; he had already been on probation for fighting. In the end, Tyler was suspended for a year and his credits for the semester nullified. Megan is pretty sure he won't be back, though she can't say whether he's learned anything from the experience. "After the hearing he said he was sorry I felt the way I did, but he never apologized," she said. "He never believed he'd done anything wrong." In fact, she confessed, *she* had to control herself from apologizing to *him*. "I hated him," she said, "but it was weird. I also wanted to give him a hug and tell him I was sorry for doing all this, for ruining his life."

DESPICABLE ME PLAYED on TV at an off-campus house as Megan and her friends poured pregame shots into candy-colored glasses. There were six girls and two boys, who were in town visiting from another school. They traded war stories about hangovers they'd had, the hazards of Everclear, and the crazy drinks they had tried: Jungle Juice, apple pie moonshine, vodka infused with cannabis or Skittles candy. Over the next hour, Megan and the other girls in the group would knock back four or five shots each. The boys would drink six. "We have a system," one of the boys told me. "Drink three shots, wait three minutes, drink two more shots, wait five minutes, one more shot and you're done." I asked what the wait was for. "So we can have time to see if we're too affected by it," he said, apparently in all seriousness.

In between drinks, the group chatted, texted friends, and posted selfies to Instagram, always looking carefully around to make sure no liquor was visible in the frame (all of them were underage). "It doesn't happen unless it happens on Instagram!" Megan told me, only half-joking. Each of them had a few stock expressions they could call up on command: a sexy chin drop, a "this is my friend and I love her" smile, an open-mouthed "aren't I crazy and having fun" face. The boys clowned around, striking the classic "sorority squat" pose. One of them checked his feed. "I only have one 'like,'" he complained. "By now I should have forty-seven!" They spent at least half their time together engrossed in their individual screens.

I doubt they realized how often they referenced gender, whether it was when a boy called a female high school classmate "all Christian during the day and slutty at night," or during a good-natured argument over which sex ultimately pays more for a frat party: the brothers, who buy all the liquor, or the girls, who have the "upkeep" of hair, nails, clothes, shoes, and makeup.

The girls reminded the boys that their cost wasn't only monetary. "Like, we have to remove our hair *everywhere*," one said.

"No razors below the neck for me," answered a boy, laughing.

"And okay," said another girl, "we have to walk in five-inch heels."

At that, the boys conceded. The girls had won, if you call that winning.

They talked, too, about the collateral damage of the party scene: a girl they knew who was bulimic; another who was in rehab; the frats that had been kicked off campus; the drunk boy who tried, with tragic results, to do a backflip off a bar.

The song "Blurred Lines" came up on the playlist, with its hooky, contentious chorus, "I know you want it, I know you want it." Megan bobbed her head in time to the beat, seemingly indifferent to the lyrics.

Surprisingly, Megan said, after the rape, her sex drive became even stronger. Like Holly, she didn't want a negative experience to define her or her college years. "I had lots of casual sex for a while. And it was good. I liked feeling giddy in the morning again instead of horrible, like when I left Tyler." But now, in the second semester of her sophomore year, she was growing weary of one-night stands. "I do get hurt feelings a lot," she said. "I set myself up for it. I know it's going to end without him texting. It does every time. Guys don't respect you after they have sex with you. That's just how it is. And that sucks. You do want that text, though. I mean just someone saying that was fun and we should hang out. If someone doesn't text me for three days afterward, I'm like, fuck you, but then if they text you suddenly on Saturday night and say, 'Hey, wanna come over?' you feel kind of obligated because you do want to see them and that's the only way."

Even though most girls and boys claim they're generally

happy with their last hookup, majorities of both also express having had, at some point, regret over casual sex. When they do, boys tend to feel remorse about "using" someone; girls feel bad about being "used." I commented to a sophomore at a private New England college that a text seemed a pretty low standard for common decency after a night in bed with someone. "And even *that* seems like such a concession to guys," she agreed. "Meanwhile, the girl has to sit and wait. And that is torturous. If you texted first, it would freak him out. On our campus, we only have one dining hall, so there's this whole thing of seeing him and he hasn't texted and, you know, 'Look me in the eye. I don't want to marry you.' Or maybe the boy next to you in Bio has seen your boobs and now wants nothing to do with you. So it's better not to hook up with people from your classes. You don't go for someone who lives on your floor. You keep your social self and your academic self detached."

Victims or Victors?

A week after her blackout, Holly hooked up again with Robert, the boy she'd started seeing at the end of the previous semester, and the two finally had sex. It was amazing. "I woke up the next morning happy that I had sex with someone I wasn't in a relationship with, who I know and like as a person, who is a sweet guy," she said. "We were able to enjoy ourselves, experiment—and we both had orgasms. We've agreed we want to keep this casual. If there's anything going on with us, it's 'friends with benefits.' We are definitely friends. Maybe if it continues, perhaps I'll want it to be something more. But that's an 'if,' because this is all new." Looking back, Holly couldn't believe how far she'd come.

Only a year ago, she was a virgin. Only a year ago, she would have said she'd need to be in a committed relationship for at least six months before she'd have intercourse. "That's clearly changed," she said. "I've pushed the line, pushed the line, pushed the line. But it's interesting where it's taken me. I don't know if it's the culture around me that tells me my behavior is okay, so therefore I'm fine with it, or if it's because I'm older and more mature and have grown as a person." She shook her head, incredulous. "It's been such a strange journey."

The girls I met often talked about "friends with benefits" as the Holy Grail of romantic arrangements: regular sex with a caring-enough partner who makes no emotional demands of you. The truth was, though, that it could be a tricky balance to strike. " 'Friends with benefits' is something college students say they want," said Lisa Wade, the sociologist, "and maybe for good reason—it might be a very functional way to go. But that's theoretical. I don't see it happening." Among the students she followed, neither the "benefits" nor the friendships could be maintained. "The problem is, friendliness is off the script in hookup culture. The minute someone says, 'I like you,' it's interpreted as wanting a relationship. If you can't tell someone you like them as a person, then you can't really be friends, can you? So the only way to maintain an ongoing sexual relationship is to treat the other person badly, to be a jerk, so they know it's not a romantic thing." The less enthusiastic partner in those FWB encounters was not necessarily the boy. "I had two FWB situations in the past year," said one college freshman I met. "Each time, I told the guy I don't want a boyfriend right now. One kind of sputtered out without being discussed, but in one case he got more attached. He said, 'I kind of want something more,' and I was like"—she shrugged—" 'I kinda don't.' I liked him. It was fun to spend time together, and I was attracted to him, but in the end, I didn't like him *enough*. That's what it comes down to. And now we're not friends anymore, really, which sucks."

Holly and Robert continued their . . . whatever it was, through the fall and winter of her sophomore year. But in March, when I checked in with her one last time via Skype, he had just broken it off. Holly, it turned out, had "caught feelings" for him and initiated "the Talk," to DTR (define the relationship). He wasn't interested. They hooked up one last time, on St. Patrick's Day, when she was "incredibly intoxicated." She described lying on top of him, naked from the waist down, and leaning in for a kiss; he turned his face away and said no. That had hurt. "I'll say it," she told me. "I definitely loved him, and the times he and I spent together were some of the happiest I have spent this year. To be honest, right now I feel like complete shit. But I want to make it clear: I don't regret any of this nonrelationship. Even though we were never officially boyfriend and girlfriend, we had feelings for one another, cared for one another, and enjoyed ourselves together. So, though in many respects this is the classic example of the way the hookup culture has 'damaged' relationships, I want you to know: I am not a victim of that culture, but a participant in it."

BY ELEVEN O'CLOCK, the streets around Megan's campus were crowded with girls in tiny skirts, boys hoisting beers. It was the first weekend of spring, and everyone was partying. As we cut through a quiet quad, a couple of boys called out, "Come here!" to Megan. When she didn't respond, they yelled, "Where are *you* going!" Then, still met with rejection, they sneered, "*Sluts!*"

"I hate that," Megan said, rolling her eyes.

Like Holly, Megan tended to blame herself rather than a

persistent double standard when she was treated disrespectfully. "Boys don't take me seriously," she'd told me. "I kind of ruined that. I sabotaged myself. I try to meet new people and go to parties where I can be seen differently. If they find out about me, they feel like they have more leeway to grab my ass or try to make out with me on the dance floor. No one wants to take the slutty girl on a date. It bothers me, but not enough for me to change my behavior."

Leslie Bell, the psychologist and the author of *Hard to Get*, has said that women are neither "primarily victims nor victors in hookup culture, but they are often misinformed." They need, she believes, to clearly understand what they can and will not get out of casual encounters—hookups are unlikely, for instance, to help them develop the skills necessary to have either good sex or good relationships. That's wise advice, but it doesn't change the terms of the debate. Some girls bragged to me that they could "have sex like a guy," by which they meant they could engage without emotion, they could objectify their partners as fully and reductively as boys often objectified them. That seemed a sad, low road to equality. What if, instead, they expected boys to be as sexually giving as girls? What if they were taught that all sexual partners, whether total strangers or intimates, deserved esteem and generosity, just as people do in any human interaction? What if they refused to settle for anything less?

It was time for me to return to the land of the grownups— Megan was heading to a frat party, and we both knew that I'd never get past the bouncer at the front door. Megan fussed over me, worrying about whether I could find my way across campus alone, where I'd get a cab, whether I'd be all right. We said goodbye and hugged, and I began to walk away.

"Be safe!" Megan called after me.

And I thought to myself, "You too."

CHAPTER 5

Out: Online and IRL

The snow had been falling thick and wet all morning outside my hotel room window. Two inches. Four inches. Six. By two o'clock, everything in the midwestern college town I was visiting had shut down. Classes had been called off. No cars or buses braved the slick roads. Students from the ski and snowboard club had jury-rigged a speaker system at the top of a hill that was barely steep enough for a child to sled down, and they were giddily, somewhat tipsily, freestyling to their beats. By three o'clock dusk was settling, and all my appointments for the day had been cancelled.

Except for one. Far down the street, I spotted a figure trudging in Timberland boots and a down jacket, hands jammed into pockets, shoulders hunched against the wind. I headed to the lobby, arriving just in time to catch a blast of cold air as the revolving door spun. There was the stomping of snow-covered boots, the unwrapping of a scarf from pink cheeks, the doffing of gloves. A hand extended to give mine a firm-gripped shake. "You must be Peggy." A smile, a look squarely into my eyes. "I'm Amber McNeill."

The New Street Corner

I shouldn't have been surprised that Amber braved a blizzard to meet me. The gay girls who responded to my e-mail queries were the most insistent about being heard. "I am a young, queer woman of color," one girl wrote to me. "We *have* to talk—I am your unicorn!" I received more responses than I expected from queer girls across spectrums of both ethnicity and orientation. One eighteen-year-old Korean American identified as asexual: not physically attracted to either men or women. I have to admit, that one threw me—interviewing her, I felt like I was talking to a lifelong vegan for a book on the joys of eating meat. But she wanted it on the record that hers was a legitimate sexual orientation, not arising from abuse or rejection. "I don't recall ever feeling any other way," she said. "I was just never interested in sex. I find it kind of . . . gross." What's more, she added, there is a thriving asexual community on the Internet: support groups, educational material, meetup sites.

At the beginning of every interview I conducted, I asked which pronoun—or combination of pronouns—to use when referring to a girl's sexual partners. Many identified unambiguously as straight or gay, others as bisexual or bi-curious. Several times an interview itself became a place to explore incipient feelings. Lizzy, for instance, a soft-spoken eighteen-year-old in the first month of her freshman year at a mid-Atlantic college, fidgeted and blushed through much of our discussion, staring at the floor or past my shoulder as she spoke. A miasma of low-grade depression seemed to hover around her, and she was so unresponsive that I began to wonder why she had volunteered to talk to me at all. She told me she had been the type of girl who was excluded and bullied in high school, called "bitch" and "fat" by the "athletic-pretty-smart 'whole package' girls that boys generally like." Still, she did have a boyfriend during her junior year, a fellow clarinetist in the school orchestra named Will. "I never really felt sexual desire for him, though," she said. "It was more like he was my best friend. We would hang out, watch TV, go to the movies. Sometimes we'd kiss a little bit, but not full-on making out."

I asked her what those sessions felt like. She shrugged. "Nice, I guess. It wasn't really my thing. To be honest, I don't really understand what's so great about it." After about four months, Will began to push her to go further—much further—via increasingly insistent texts: "We should totally have sex!" he wrote, and "Come on! It will be fun! It will be great!" and "Why not? I don't understand!" "I told him he was making me uncomfortable," Lizzy said. "We'd never even done anything below the neck! But he would just keep bringing it up, texting me over and over."

Although Lizzy didn't think she should have to justify a disinterest in intercourse with a boy she'd barely kissed, one who demonstrably had no respect for her limits and whose conversational skills did not extend past the keyboard, she nonetheless tried. Maybe, she said, her reluctance stemmed from shame over her body. "You see a lot of models and superstars, and they're so skinny and gorgeous," she said, looking down at her soft belly. "Even shopping for clothes—clothes are cut for people who are skinny, and I'm just not skinny." Then she shook her head. "But really, I wasn't attracted to him enough to even want to try. It was just, 'Oh, no! He wants to have sex and I don't.'" After two months of fending him off, she suggested they "take a break." Will, her supposed "best friend," never spoke to her again.

Other boys, and even adult men, had shown interest in her since, but she never reciprocated; the prospect of physical intimacy repelled her. I asked her to recall a time when she felt sexual pleasure in her body. She blushed. "I can't think of one," she said. What about arousal? The color in her cheeks deepened. "I haven't explored any of that. I just want to get through my classes and do my work. And it's hard to open up to people. It takes a lot of effort."

I could see that for her it did—our conversation proceeded in fits and starts; she was perhaps the least voluble of the girls I met. Then I asked, "We've really only discussed boys. Have you ever felt attraction to other girls?" Again, Lizzy's face grew pink, but this time it seemed to be with pleasure. "I have this really good friend," she admitted, and then, for the first time in our conversation, she laughed. "I kind of like her both ways, you know? It's like I'm balanced on the edge. There's just something . . . amazing about her." She laughed again, her smile lighting up her face. "I can't even put my finger on it. I've never met a person where I've felt . . . it's just *there*."

Lizzy had never personally known anyone who was gay, but she'd read about homosexuality on the Internet, specifically in fan fiction: original stories penned and passed around online by devotees of popular books, TV shows, plays, movies, or pop songs. The erotic novel *Fifty Shades of Grey* famously started out as fan fiction based on *Twilight*. Harry Potter has eighty thousand fanfic stories on one site alone. A fan fiction story based on *The Hunger Games* had, at this writing, over two million views. Fan fiction may "cross over" between worlds or genres, so Harry Styles, for instance, might lose his Direction and find himself in Middle Earth. It also often includes erotic, typically same-sex, canoodling (presumably) never dreamed of by the characters' creators: Mr. Spock gets with Captain Kirk; Holmes with Watson; Batman with the Joker; Hermione with Ginny. Women and girls are the largest creators and consumers of fan fiction. It's hard to say why, then, the overwhelming percentage of its sexual encounters are between men. Maybe it's because women are still underrepresented in mainstream media, and so are less compelling as characters. Or maybe writing about male bodies liberates women from the judgments about appearance, behavior, or assertiveness that typically freight their sexual exploration. Whatever the reason, fan fiction provides a form of freedom to young women: it's generally without commercial motive or viability, a corner of the media from which, with few exceptions, no one is profiting.

Like anything in the boundless olio of the Internet, that breadth can have drawbacks as well as advantages: one eighteen-year-old girl from Staten Island recalled stumbling on graphic fan fiction in middle school. "Little girls and big girls and occasionally guys write a *lot* of porn based on characters they like," she said. "I would read it all. I didn't know about BDSM until I read fan fiction; it's in a lot of the sex scenes. And for a long time, I thought the average size of a limp penis was eight inches—and that *then* they grow larger. And I thought, 'I never want one of those near me!'"

Lizzy, an avid fan of the TV show *Dr. Who*, was first exposed to lesbianism by chance, on a Tumblr blog that coupled two characters who, in the show itself, are straight. "At first it was weird," Lizzy said, "but the actual story was really good. It worked. So I kept reading. And it broadened my view of the world. I mean, I hadn't really thought about this stuff before. It was . . . not embarrassing. Just strange. Foreign. Exciting."

Adults, me included, often fret over the hazards of the Internet for kids, especially where sex is concerned. Our fears are understandable: the easy access to extreme porn, the distorted female bodies, the sexting scandals: it is enough to make anyone born before 1980 feel that Armageddon is nigh. But as with so much of contemporary culture, it is hardly that simple. As long as adults still avoid open discussion of sexuality, teens will inevitably seek information on today's electronic street corner. That presents both problem and opportunity. Yes, there are discussion board sites such as Reddit, which can quickly devolve into creepshots of women's cleavage or teen girls' butts in short shorts, bikinis, and the like. (The company's policy against posting nonconsensual porn, announced in early 2015, has so far done nothing to abate such "communities.") But there are also Scarleteen, Go Ask Alice!, and Sex, Etc., where the advice offered may be explicit but is scrupulously medically accurate.

As with their straight peers, the Internet can be double-edged for LGBTQ teens. According to the Gay, Lesbian, and Straight Education Network's 2013 report *Out Online*, they experience cyberbullying at three times the rate of heterosexuals—girls more often than boys. Yet LGBTQ kids also turn to the Web for information and support—crucial for a population whose attempted suicide rate is still five times that of other teens. Over half of LGBTQ young people who were not out in person used the Internet to connect virtually with others like them, according to the report. More than one in ten disclosed their sexual identity to someone online before telling anyone in the "real" world, and over a quarter were more out on the Internet than they were in their offline lives.

Ideally, queer teens wouldn't need to resort to trolling gay chat rooms for information or acceptance. At the same time, the Internet has provided an unprecedented pathway to normalizing and embracing sexual identity. Lizzy offered a glimmer of how that might begin, as did the young woman who'd found online support for her asexuality. But it was nineteen-year-old Amber, at a college hundreds of miles from Lizzy's, who best illustrated the potential (and a little of the weirdness) of our hyperconnected world.

After introducing ourselves in the chilly hotel lobby, we headed up to my room; Amber settled into a wingback chair under a circle of lamplight and began to tell me how, even while keeping up the appearance of the straight, popular girl her parents expected her to be, she was secretly working through something else online, something she didn't always understand, building a second identity that, in the end, proved the most real of all.

Playing the Straight Girl

The first time Amber misrepresented herself online, she was just nine years old, doing exactly the sort of thing parents fear: chatting with strangers on a gaming site. "People would try to start these sexual conversations with me," she said. "I don't even know if I really knew what sex was. I was just a naïve kid." Eventually her parents wondered why she was spending so much time on the computer and checked her history. When they discovered what she'd been doing they instantly forbade her, indefinitely, from going online. Amber didn't mind the punishment so much as her parents' horrified reaction. "I felt like I'd been doing something really, really bad," she recalled. "I was a wreck. I didn't touch a keyboard again for a year."

When she did, though, she got into *Second Life* and *The Sims*, virtual worlds in which users, represented by onscreen avatars, can once again interact with one another. Whether on the Internet or a PlayStation, Amber always chose to be male. "I didn't think anything of it," she explained. "It was just what I liked. I would make my boy avatar, then go on these websites and talk to girls, tell them they were pretty or whatever a fifth-grader would say. I never really questioned it. I honestly didn't even know what the word *gay* meant. Nobody talked about it: not my parents, not my school. Which is weird because it's not like I grew up in the middle of nowhere: we lived near a big university. I went to a high school of three thousand students. But no one said anything. So I never questioned my sexuality."

This was, of course, years before the Supreme Court ruling that made same-sex marriage legal in all fifty states. Still, it wasn't exactly the Dark Ages: celebrities such as Melissa Etheridge and Ellen DeGeneres publicly embraced their sexuality in the 1990s. Openly gay characters were increasingly common (and nuanced) on TV and in movies, too. Perhaps as a result, the average age of coming out in the United States began to plummet: from twenty-five in 1991 to between fourteen and sixteen today. "Children report awareness of sexual attraction at about age ten," Caitlin Ryan, director of the Family Acceptance Project at San Francisco State University, told me. "That's earlier than most adults, including parents, believe. But sexual orientation isn't only about sex. It's also about social and emotional relatedness, human relationships, feelings of connection." As an example, she pointed to the Broadway musical *Fun Home*, based on cartoonist Alison Bechdel's graphic memoir. Nine-year-old "Small Alison" first confronts her own difference when she sees a butch deliverywoman enter a diner. "Ring of Keys," the show-stopping song she sings, is not about eroticism but identity, recognition: a paean to the woman's "swagger" and "bearing," her "just right" cropped hair, jeans, and lace-up boots, her way of being, of presenting to the world.

Maybe Amber's avatars were her "ring of keys." At any rate, it didn't last—once again, her parents checked her computer history and discovered what she was up to. By then they had divorced, and her father had moved out of state. Amber remembers an airport handoff, sitting in her mom's car while her parents conferred on the curb. "It happened again," she heard her dad say grimly. Later, her mother asked Amber why she'd chosen male avatars, but before Amber could answer, her mother fed her the response she wanted to hear. "She said, 'You just wanted to see what it was like, right?'" Amber recalled. "And I was like, 'Yeah, yeah, that's right—I wanted to see what it was like.'" If her mother harbored any ideas about her daughter's sexuality, she didn't let Amber know.

It doesn't take long for kids to exceed their parents' competence online. By eighth grade, Amber was savvy enough to erase her browsing history, create untraceable free e-mail accounts, cover her tracks. Posing as a boy named Jake, she built a fake MySpace page, posting a profile picture she'd downloaded of a cute guy from her school and claiming to be from Los Angeles. If you'd asked her at the time, she wouldn't have been able to say why she was doing it; only in retrospect can she connect her behavior to her sexual orientation. For two years she used the page as a cover to flirt with what she described as "oodles and oodles" of girls. None of them ever caught on, even when they spoke to her on the phone. (Amber demonstrated her quite credible imitation of a teenage boy's voice.) She did make one mistake, though: she gave them her real cell phone number, attributing its midwestern area code to a recent move. That was six years ago: she still gets texts from some of those girls. "I got one the other day out of the clear blue sky that said, 'I miss you,'" she said. "It's sort of weird."

It occurred to me that perhaps the ideal imaginary boyfriend for a teenage girl might very well be another girl pretending to be a guy. Who would better know what she wanted to hear? Amber agreed. "I think they look back to when they were in high school and think, 'Oh, I remember that one guy: he was so nice, and he always really *understood*.'"

Recalling that period herself, though, brings Amber pain. She feels ashamed and guilty about deceiving other girls. "It bothered me for a long time," she said. "I'm mostly over it now, but then I get these text messages and I'm like, *What the hell?* They come out of the woodwork. You'd think after you've watched enough *Catfish* episodes, you'd realize that I probably wasn't actually a real person with the wrong area code.

"It's sort of sad," she added, "when you think about it."

If Amber was impersonating a boy online, in real life she was learning, after a fashion, to impersonate a girl—or at least a certain kind of girl. Up until puberty, Amber passed as a "tomboy." She wore loose clothes, slicked back her hair, sometimes pretended to shave with her dad. If she was occasionally mistaken for male? Well, that was fine with her. No one *forced* her to change, exactly, but as she hit middle school, the expectation was clear. Her mom had been a cheerleader in high school; her dad is an orthodontist. Appearances mattered to both of them. Maybe her parents had their suspicions about Amber's sexual orientation; maybe they were hoping to stifle it. At the very least they were eager to have her behave like a conventionally feminine girl. They encouraged her to wear skirts, and her mom taught Amber to apply makeup. "I didn't want to be the 'weird' kid," Amber recalled. "So I just had to, you know, go with the flow. I'd wear mascara and I would say, 'Oh yeah, I *love* Zac Efron!' because I wanted to fit in. But I was always tugging on the clothes, I never felt comfortable. I was just going with the flow—I was always going with the flow."

Amber tried to join in as her friends experimented with "relationships" that would last a week or so, but whenever a boy put his arm around her, she pushed it off. "I'd tell my friends he was weird or creepy or clingy," she said. "Then I'd ask them to 'break up' with him for me." At fifteen, though, Amber met a boy who, coincidentally, was named Jake. She was drawn to him immediately. "We were best friends. My mom used to say we were like the same person in different bodies. We would play video games and watch movies. I would hang out with his family; he would hang out with mine." She didn't encourage a romance, but, she said, as with the miniskirts and lip gloss, "I was going along with things, so why not go along with having a boyfriend, too?"

To Amber's relief, Jake was a devout Christian who planned to remain a virgin until marriage. So, she figured, she had "nothing to worry about." For a few months, through the fall of her sophomore year, the couple did little more than kiss. While Amber didn't enjoy it, neither was she unwilling. "I never really had any feelings when we were doing it," she recalled. "It didn't turn me on. It just . . . happened."

In January, Jake invited her to their school's winter formal. She said yes, though the idea of grinding on the dance floor in a short skirt was not appealing. She found a knee-length red dress that, as she said, "was edgy, but didn't show any boob" and wore stiletto heels ("but not strappy," she said; "and they had a closed toe"). As for the dancing? She tolerated it—which, truth be told, was the case for many of the straight girls I spoke to as well. Afterward, Jake suggested they grab a soda at a McDonald's drive-thru, and then sit and talk in the car for a while. Amber agreed. "I'm thinking, it's me and Jake, you know?" she said. "So, fine, whatever." They pulled into a church parking lot. Jake turned off the motor and leaned in for a kiss. Then, without warning, he slid his hand under Amber's skirt. She broke out in a cold sweat and her stomach clenched, but she remained silent. When he suggested they move to the backseat, Amber, yet again, "went with the flow."

She went with the flow as Jake took her hand and shoved it into his pants. She went with the flow as he slid her underwear aside. "Then," Amber said, "God, he was only a sixteen-year-old boy—his finger goes in the wrong place. It goes up my butt hole!" Jake was mortified. "I'm so sorry! I'm so sorry!" he repeated. Amber assured him she was fine—she didn't want him to feel bad, she said—but the mood, such as it was, was shattered. He zipped his pants and slunk into the front seat. "It was actually the best thing that could've happened," Amber said now. "Because it ended things. He just drove me home, and I was like, '*Yes*! It's over!'"

Though of course it wasn't. Since she'd allowed him to touch her once, Jake assumed he could do it again. And Amber never did say no. She also never said yes, and he interpreted her passivity as consent. She would sit unmoving, hands at her side, staring into space as he groped and rubbed against her. "Once he asked why I didn't make the same noises as girls in porn videos," she said. "He watched a *lot* of porn. I told him I was quiet because I was so into it. So, he thought I liked it. He thought it was normal, and I let him think that. Because I was go-with-the-flow Amber."

Most of the gay and bisexual girls I met had gone through a period of trying to pass as straight, sometimes experimenting with lesbianism under cover of heterosexuality. A bisexual high school senior in San Francisco, for instance, would go to an all-ages club so she could make out with other girls on the dance floor. "They were doing it mostly to get attention from boys," she recalled. "Whereas I wasn't. But they didn't know that. So it was really great." Later, she went further, bringing a second girl into bed with her boyfriend; by her freshman year of college, she was dating a woman. In general, girls have become more open to same-sex attraction in recent years, more accepting of sexual fluidity. In the early 1990s, for instance, only 3 percent of women who identified as heterosexual in *The Sex Lives of College Students* reported some same-sex experience; by 2008 nearly a third did (though, again, no distinction was drawn between girl-on-girl action performed mainly to titillate guys and the real thing).

For Amber, flowing with the hetero current became increasingly difficult. She knew she did not—*could* not—feel about Jake, or any boy, the way her friends did. "They would pull out pictures of guys they met over the summer or on Facebook and be like, 'Oh, he's so hot, I just want him to fuck me,'" Amber said. "And I'd be like, 'Um, *yeah*, me, too.' That was all I could say. Or sometimes: 'He's really attractive.' I never said a guy was hot or even good-looking. I never thought any of them were."

Like Lizzy, Amber had not to her knowledge personally met a lesbian, though she had seen them on TV shows such as *The L Word*. She worried that her feelings wouldn't be seen as normal, that she would embarrass her mother, disappoint her father, alienate her friends. By the fall of her junior year of high school, the effort of keeping up the straight-girl facade was leaving her exhausted and depressed. So she turned to the only outlet she could think of: the Internet. "I needed to find someone to vent to," Amber recalled. "I thought I would release it all and that would be enough; I'd be able to suppress it again for a few more years." She searched Tumblr for gay blogs, something I tried myself and that, at least initially, returned an array of photos of men: some kissing sweetly; others naked, stroking outsize erections; ejaculating onto one another's faces; performing oral or anal sex in duos, trios, or larger groups. The results for "lesbian" were equally graphic, though adding "teen" pulled up, along with the XXX fare, a smattering of angsty quotes, pictures of dancing cats, and carefully curated selfies. On a page called "Girls Who Like Girls," Amber stumbled on Hannah, who was squarely in the nonexplicit, angsty quote camp. Hannah posted her own writing as well as pictures of places she dreamed of visiting in Paris, London, and Rome. There were no photos of her face, Amber recalled. "That made it seem like she really just wanted to talk." She also lived far away, in Ottawa, Canada. "It was perfect. I was going to vent to her about all the fucked-up things that I'd done, and then I would never talk to her again."

Amber paused, shaking her head. "So wrong," she continued. "*So* wrong.

"Hannah rocked my world."

Coming Out in the Twenty-First Century

On another winter evening, a few months after we first met, Amber introduced me to Hannah. They were nearly three thousand miles and an international border away from my California home, but thanks to Skype, we were all in the same room. Hannah jumped up every few minutes to check on a chocolate chip banana bread she was baking for Amber. ("It's her favorite," she explained.) They talked about the party they went to on New Year's Eve; about how, for Christmas, Amber took Hannah ice skating and gave her a necklace; about the last time they were together with their families. They sat close, draping their arms around each other, touching constantly in the way of young lovers. Amber wore a hoodie from Hannah's university; Hannah wore a T-shirt from Amber's college, her long dark hair covering the school's insignia.

Five minutes after Amber sent that first, fateful message, she got a reply from Hannah suggesting they Skype. They did, and ended up talking until four in the morning. "I told her everything," Amber said now, gazing at Hannah affectionately. "About the fake MySpace profile, about getting caught by my parents, everything. It was crazy. I knew in, like, a split second that I didn't want to talk to anybody else ever for the rest of my life. She was the first person to tell me my feelings were okay. And I realized: *this* is what a relationship is supposed to feel like. You're supposed to feel appreciated and accepted and comfortable and able to say anything." Hannah's eyes welled up, and Amber pulled her close. "Why are you crying?" she asked.

"Because you were so sad," Hannah replied. "You needed someone to listen to you. I remember thinking, 'This girl really needs someone to tell her it's okay.'"

Within a few weeks, Amber's relationship with Jake fizzled, and they agreed to split up. While she was now free, she was only sixteen, and the new object of her affection lived in Canada. There was no way Amber could see Hannah in person—not, at least, without coming clean to her parents.

YouTube is full of "coming out videos"—that phrase returns about twenty-one million results. There are poignant and funny videos, and some that are heartrending, as parents accept or reject their children live on-screen. There are videos of twins coming out together. There is a subgenre of "how to come out" videos, and another of songs people have written about coming out. Amber watched dozens of them, trying to get up her nerve to talk to her mother. She resolved to do it over winter break, but as Christmas turned to New Year's, she continued to put off the conversation. Finally, just before school restarted, she invited her mother out to lunch, not something she typically did. It was a strategic choice: her mom wouldn't make a scene, Amber figured, in a public place. They agreed to meet at a deli. Amber was so nervous that morning she was shaking: she still wonders how she drove there without crashing the car. Her mother was already at a table, looking stricken. "Are you pregnant?" she burst out, before Amber was even seated. Amber laughed, and said, "No, Mom," thinking to herself, "The farthest thing from it."

Amber unfolded a piece of paper, a letter she'd written a month earlier and carried with her ever since. "I love you and I don't want to disappoint you and I always want to make you happy," she read. Then came the two words: *I'm gay.* But when she got to them, she choked. "I couldn't say it, I think because I had not accepted it myself. Finally, I don't even know how, it just came out." At first her mother seemed relieved. Her daughter wasn't on drugs. She hadn't stolen anything. She wasn't pregnant and didn't have an STD. Nor did Amber want to move in with her dad. Her mother hugged Amber, told her it was fine, just fine. "I love you," she said. Then the conversation took a turn. "How do you *know* you're gay?" she asked. "Maybe it's just a phase." Maybe, she continued, it had something to do with the divorce, with having a poor masculine role model. "There was no way in hell she was going to believe I was born this way," Amber said. "She just didn't understand."

As the age of coming out has dropped, parental support has become more crucial than ever. It's one thing for your mom and dad to banish you from the home at twenty-five; it's quite another at twelve. In a survey of more than ten thousand teenagers, those who were LGBT-identified listed tolerance and their family situation as the things they would most like to change in their lives; other kids said finances and their weight or appearance. LGBT kids also cited family as their "most important problem"; other kids said grades. According to Caitlin Ryan, family acceptance is the single biggest factor in an LGBT child's well-being. Ryan's organization has linked rejection by parents to increased risk of suicide, depression, abuse of illegal drugs, and HIV/AIDS. To a degree, this would seem self-evident. Less obvious is what teens experience as "rejection." Parental silence, for instance: one girl mentioned angrily to me that while her mother's Facebook page was plastered with pictures of her brother and his girlfriend, there was not a single photo of her and *her* girlfriend. Kids also consider the kind of comments Amber's mom made ("Are you sure?" or "Maybe it's a phase") as profoundly hurtful. Letting insulting comments by extended family slide is right up there as well. That said, Ryan found that most of those negative or ambivalent responses come from a place of love. "Parents are often expressing fear and anxiety exacerbated by misinformation," she said. "They wonder: 'What's going to happen to my child in the world? How do I deal with this in my own family? How do I reconcile conflicting beliefs?' The good news is that a little change in their response can make a huge difference."

It would take months of arguments and tension for Amber's mother to come around; it certainly did not seem the right time to tell her about Hannah. So Amber put that part of the conversation off, and then put it off some more. Meanwhile, the two girls continued to Skype late into the night.

"Who are you always talking to?" Amber's little sister would ask.

Amber would shrug. "A friend," she'd respond.

That did not satisfy the younger girl's curiosity. She began to grow suspicious, hostile. There was the time she snuck up behind Amber in the family's laundry room and said, "You're a fag, Amber. You're a dyke." Other times she would hiss, "You're such a *lesbian.*"

"I think my sister just didn't know a healthy way to get me to come out to her, and she really wanted to know," Amber said. When I admired her generosity of spirit, she added, "Well, it definitely hurt my feelings. And even today it puts a damper on our relationship. I mean, who does that to their sister?"

Coming out, of course, is not a one-time deal. A person has to do it over and over, not only to people she already knows, but to everyone she will ever meet. Amber tried confiding in a few trusted friends, breaking the news, perhaps unsurprisingly, over Facebook chat or text. "I could never do it in person," she said. The girls always assured her that nothing had changed, but they wouldn't mention the conversation again, and invariably the friendships drifted. "I would think, 'All right, I guess you're just busy.' But, looking back, I realize they didn't want to be friends with me anymore once they knew I was gay."

Amber never did work up the nerve to tell her mom about Hannah—not directly. One day, though, while she was in her room, her mother stormed in, pushing the door open so hard that it banged against the wall. She brandished Amber's cell phone. "Who the hell is *this*?" she yelled.

She had read all Amber's and Hannah's texts, including the ones in which the two declared their love for each other. Amber just stared. "It was the worst possible way she could have found out," she recalled. "It was so upsetting. Because if she wasn't really accepting about my being gay, she sure as hell was not going to accept a long-distance relationship with some random girl."

"What do you think you're doing?" her mother continued. "How old is this person? How do you know she's not thirty-five?"

Hannah's mother—her father had passed away—was more accepting of her daughter's sexuality and budding relationship: she offered to try to smooth things over with Amber's mom and arrange for the girls to visit. Amber's mother refused. No way was her daughter visiting some stranger in Canada. "I would *beg* her to let me see Hannah, just for a day," Amber recalled. "It didn't have to be in Canada: her mom would let her visit us. But my mom said no. It was like she thought if she kept us apart, I wouldn't be gay anymore."

By summer, Amber's mother had calmed down enough to allow Hannah to visit for three days. She could stay at their house, but the girls had to sleep two levels apart. Amber didn't care. She was going to see her girlfriend in person. Nervously, she Googled "what do two girls do together." She needn't have worried: the instant they were alone, Amber and Hannah began kissing, and the feeling was unlike anything Amber had previously experienced. "I was so into it," she recalled. "It was just a natural, normal thing, exactly how it was supposed to be. That's how it probably is for everybody else when they're in an intimate relationship, that kind of feeling I had."

Girls in relationships with other girls spoke very differently about sex from those who were involved with boys. A senior at a California public high school who identified as bisexual told me that she enjoyed the reciprocity she found—and had found *only*—in her same-sex encounters. "It's so different," she explained. "It's like my turn, her turn, my turn, her turn." Another bisexual high school senior said she tended to be more passive with male partners. "With another girl . . . well, you can't *both* be passive. Nothing would happen. With a guy it feels like he's doing something *to* you, but with a girl, you're doing it *with* each other." A college junior in the Midwest told me that sex with her girlfriend felt "off the script": since there was nothing they were *supposed* to do, they were free to create the sex life that worked for them.

Because she had never had intercourse with Jake, Amber considered herself a virgin when she and Hannah met. I asked her if she believed herself to be one now. She shook her head. "I was so confused, though, so unclear about what 'gay' meant, that I had to Google 'When is a lesbian not a virgin?'"

What was the answer? I asked her. "There wasn't one," Amber said. "For me . . ." She paused for a long moment. "I think it's just the second that you are being intimate, touching each other more than just kissing. Not your breasts, necessarily, but below the waist. The second you touch there, you're not a virgin anymore.

"But honestly, I don't really have a definition. I just knew. I guess I would define it . . . maybe once you have an orgasm with someone? Once you have an orgasm with someone, you're definitely not a virgin anymore. Yeah, that's how I would define it."

When Is a Girl Not a Girl?

Amber and Hannah's relationship deepened during their senior year of high school, and as it did, Amber grew more confident in other parts of her life. She discovered she liked public speaking and acted as emcee for a school-wide talent show; she was elected to the homecoming court; she socialized more. Although she stayed in the closet for the most part, she ditched the skirts and makeup in favor of a more pared-down, androgynous look. "I would just own it," she said. "It was fun! And nobody had a problem with it."

That first afternoon we met, she pulled up a photo on her phone of what she used to look like and passed it over to me. The girl in the photograph—her blond-streaked, carefully styled hair flowing to her shoulders, candy-pink lipstick, blue eyeliner and mascara—looked nothing like the young woman in front of me. At the same time, the current version of Amber wasn't much different from any of the straight girls I'd met, at least when they were dressed for school rather than for the party scene: she wore jeans—she said they were men's, but I wouldn't have guessed—a hoodie emblazoned with her college's name, and hiking boots. She wore no makeup, but neither, during the day, did many other girls. Her hair was pulled back with a black headband and fastened into a short ponytail. Even so, during our first conversation, the planes and shadows of her face seemed perpetually to shift: maybe it was a trick of the light, or maybe I was just tired, but sometimes she looked clearly like a girl and other times, quite suddenly, she could easily have passed for a boy.

Amber was not always certain herself which she was. It was in trying to answer that question that she found the Internet, previously so dependable, finally failing her. YouTube videos and websites she scoured suggested she might be transgender, a term she had never before heard. (It would be years before Laverne Cox and Caitlyn Jenner graced the covers of glossy magazines.) She spent the next twelve months, until just before she left for college, worrying that it was true. "It scared the living shit out of me," she said. "I was like, what am I going to do? I was going to have to go through all these surgeries and get my name changed. I thought it was the only option."

Certainly the Internet can be a trove of misinformation, distortion, dis-expertise, and bad advice. On Google, a nicked cuticle becomes a life-threatening emergency; so does working out or taking a shower (though, if you stopped exercising, you could bathe less and minimize both risks). So a young gay woman who had never heard the word *butch*, let alone *transgender*, could easily become confused, especially if, like Amber, she was raised in a community with conventional ideas about masculinity and femininity. An estimated 0.3 percent of Americans are thought to identify as transgender—that's close to seven hundred thousand individuals. (About 3.5 percent of adults identify as gay, lesbian, or bisexual, though rates are higher among those ages eighteen to twenty-nine.) The true number is hard to quantify, though, since it may or may not include those who identify as "genderqueer"—living between genders, beyond genders, or as a combination of genders. At its fullest (and some would say most threatening) manifestation, genderqueer upends notions of femaleness and maleness, masculinity and femininity, changing them from a biological inevitability into a customizable, ever-changing buffet of identities, expressions, and preferences. There was the 2013 story, for instance, of Arin Andrews, who began life in a girl's body, and Katie Hill, who began life as a boy. They fell in love in a support group for transgender teens, went through their transitions together, and continued forward as a heterosexual couple. Or the darker tale of Sasha Fleischman, born a boy in Oakland, California, who is agender—that is, not identified with either sex, preferring to be addressed by the pronoun "they." As a high school senior, Sasha suffered severe burns to the legs when another teen set Sasha's skirt on fire aboard a city bus. An outpouring of support followed—a local protest march peopled by boys wearing "skirts for Sasha"; thousands of dollars raised on the Internet to offset medical bills; local school policies changed to allow gender-nonconforming students to choose which bathrooms and locker rooms they want to use, which sports teams they want to join.

Modern college campuses are replete with gender warriors who specify whether they are cis-gender (meaning their emotional, psychological, physiological, and genetic genders match), nonconforming, or transgender. They may replace *he* and *she* with neutral pronouns such as *ze, ne, ou, hir, they,* or even *it.* The rejection of the "gender binary" can be truly radically liberating. At the same time, a rush to label a young person as "nonconforming" may risk unwittingly calcifying traditional categories. Consider the case of a male-to-female transgender first-grader whose family sued her Colorado school for forbidding her to use the girls' bathroom: her parents said their first inkling that their son, the only boy in a set of triplets, was unusual came when he was five months old and reached for a pink blanket meant for one of his sisters. Later, he rejected a car he was given for Christmas, showed no interest in sports-themed clothing, and donned a princess dress rather than a fireman's uniform in fantasy play. Five-month-olds don't know pink from blue. And choosing tulle over tools? With all due respect to the family and the child, who may indeed be transgender, that hardly seems like "proof" of anything other than adult bias. Yet nearly every press report I read not only trotted out those anecdotes but placed them in the story's lead. Even as I admired the child's parents for supporting their daughter, that inflexible definition of masculinity—which would see a boy as actually *female* before accommodating his love for sparkly gowns—concerned me.

Some of Amber's reasons for questioning her gender identity were similarly retrograde: they included being more dominant in bed, standing up for herself, planning to pursue a career in business, and hating to cook. Nor was Amber the only young lesbian I met who wondered whether her clothing and attitudes meant she was actually male. Valentina, eighteen, the girl who called herself my "unicorn," also spent her senior year of high school thinking she "must" be transgender. Growing up in a low-income, largely Mexican American neighborhood in Chicago, she shunned anything conventionally feminine: Barbies, pink, skirts, frills. Dressed in a flannel shirt and loose jeans, she told me that in middle school other girls would crawl into her lap to cuddle, calling her "Big Daddy" (she was broadly built) and ask advice about boys. By high school, she was scouring the Internet for clues as to her identity. "I wanted to know," she said, " 'Am I gay?' 'Am I transgender?' "

"Did you feel like you were in the wrong body?" I asked her.

"No."

"Did you feel like you wanted to be a man?"

"No," she said again.

"Then why did you think you might be transgender?"

"That's exactly it!" she said. "What finally made me realize I wasn't trans was reading about people who said, 'I felt like there was a guy inside of me trying to get out.' I never felt that. I never felt like I should be a guy. I like my vagina. I wouldn't want anything to happen to it. But I wasn't sure I wanted to be a girl, either."

Such confusion is understandable, according to Jack Halberstam, professor of English and director of the Center for Feminist Research at the University of Southern California, who writes about transgender issues. Even as some young people may be helped—sometimes saved—by the recent visibility of transwomen such as Cox and Jenner or by TV shows such as Amazon's *Transparent*, tensions between "butch" women and transmen have been building. "The whole concept of 'butch' is now seen as a kind of waiting room in which you stay until you change your gender physically," Halberstam said. "We don't have words for someone who has strong cross-gender identification but feels good about their bodies. *Butch* has become anachronistic, but *trans* implies transition, possibly hormones and surgery. *Genderqueer* is as good a holding term, but it's clumsy, really." I pondered the idea of "cross-gender identification." That, too, seems often culturally determined, to the detriment of girls such as Valentina and Amber. When we've defined femininity for their generation so narrowly, in such a sexualized, commercialized, heteroeroticized way, where is the space, the vision, the celebration of other ways to be a girl?

As for Amber, "I looked heavily into it," she told me during our first meeting. "I could tell you everything under the sun about being transgender. I'd go through these lists I'd find online. They would ask questions like, 'Do you cry when you think about having a vagina?' And I'd think, 'No, not really.' Maybe if somebody told me I could choose one sex or the other I would have picked the other, but I don't feel upset about it. I had all these conflicting feelings. Like, I don't really care about my boobs. That's weird, right? So then I dealt with 'Am I a biological *mistake?*' "

Ultimately, Amber realized she did not want to give up who she was, did not want to be someone completely new: "I mean, say your name is Cheryl," she explained, "and you're becoming Sean. You have to not want to be Cheryl anymore and never talk about Cheryl again.

"Well," she added, sitting forward in her chair, "I love being Amber. I could never in a million years imagine not being Amber. *I am Amber.* And I don't know if I fit being a lesbian perfectly, but I'm definitely not a transgender person. I can live my life in this body, confident and happy, and in a healthy relationship." She leaned back again, letting her hands drop to her lap. "And it took me a year, an entire year, to be able to sit here and tell you that."

DURING OUR CONVERSATION on Skype, Amber and Hannah told me they have no fear of holding hands, snuggling, or kissing on the streets of Ottawa. They're a bit more circumspect when Hannah visits Amber. Amber's mom still hasn't fully come around. ("She'll never accept us like she would if I were with a straight person," Amber said.) And while people have been mostly accepting on her college campus, the couple has occasionally been harassed by groups of young men. Still, Amber has been experimenting with being more public about who she is. She recently volunteered to serve on a panel of LGBTQ people who visit classes at her school to talk about their experiences and answer questions. She has also declared a double major in economics and public policy and is toying with going to law school and ultimately entering politics. "I'd like to try and get a House of Representatives seat," she told me, and then laughed. "I've got a little bit against me in running for office—I'm both a girl and gay. But I'll figure it out."

"Well," I said, "it just might be your time."

She nodded, smiling. "I always tell myself that," she said. "Things are opening up for women and us gay people. I guess we'll see."

CHAPTER 6

Blurred Lines, Take Two

I met Maddie Reed at the community college where she was en-rolled in a special program for home-schooled or otherwise "independent" high schoolers. She shook my hand and smiled, a pale, curvy girl with a spray of freckles across her nose and reddish-brown hair that hung past her shoulders. She'd attended classes here for a semester and planned to stay another year, un-til graduation. That meant she wouldn't, as she once dreamed, write for her high school newspaper, or go out for the softball team, or attend the prom. "I don't think about it," she said as we strolled through campus, looking for a quiet corner where we could talk. "I don't let myself. It's still a fresh wound. And I know other girls have it way worse. At least I have a vague idea what happened to me. At least there weren't pictures of it going around. But I wasn't really aware that this was a problem before. I thought it was just something that happened in . . . I don't know, other parts of the world."

Who Stole Consent

The attention to sexual assault over the past few years has been unprecedented. From dorm rooms to press rooms to the White House, fighting rape, especially on college campuses, has become one of the most prominent, contentious civil rights issues of our time, up there with gay marriage, abortion, and police brutality. What is the definition of sexual assault? What constitutes consent? How should schools fairly handle allegations? This is not, though, the first time that acquaintance rape has sparked debate. The late 1980s and early 1990s saw a flurry of divisive, high-profile cases. The first, and perhaps most appalling, was in 1989, in Glen Ridge, New Jersey, where a group of high school boys assaulted a mentally disabled girl (a young woman they'd known since childhood) with a broomstick and a baseball bat. The case featured several elements that would resurface in the current national upheaval: the boys were star athletes in an idyllic, football-crazy town; although their actions had been initially reported as out of character for them—a "stupid mistake" by otherwise "decent" kids—they had in truth been abusing their godlike status since middle school: bullying classmates, destroying property, creating mayhem. They were disdainful of girls and female teachers (one of the boys regularly exposed himself in school and frequently masturbated during class); treated sex primarily as a form of male bonding (watching porn together, convincing younger girls to give them successive blow jobs, and secretly watching one another's escapades with unsuspecting partners). The girl they assaulted was incapacitated, though in this case mentally rather than by drugs or alcohol; bystanders declined to intervene. After the boys' arrest, many of the town's adult residents defended them, claiming that the girl was a "sexual aggressor" who had "asked for it."

Around that same time, allegations against thirty-year-old William Kennedy Smith stunned the public in a different way: he was, after all, a medical student, clean cut, wealthy, privileged—a *Kennedy*. He met his alleged victim while drinking at a Florida bar with his uncle, Senator Edward Kennedy, and his cousin Patrick, a future congressman. The woman later claimed that in the wee morning hours, Smith tackled her as they strolled on the sand near the family's Palm Beach estate; then he pinned her down and raped her. Smith insisted their sex was consensual. He was ultimately acquitted; many believed the verdict might have been different had the judge admitted testimony by three other women (a doctor, a law student, and a medical student) who, in sworn depositions, said that Smith had also assaulted them, though they didn't report those incidents to the police. Before the media had finished dissecting that case, former heavyweight champion Mike Tyson was charged and convicted in Indiana of raping an eighteen-year-old Miss Black America pageant contestant during a late-night date; he served three years of a six-year sentence. None of these assailants fit the prevailing image of the rapist as a deranged guy in a ski mask leaping out from a dark alley. The accusers knew their attackers, and had, to a point, gone with them willingly. That was, of course, used by defense lawyers as proof of consent, or at least partial collusion— the women "should have known" what was going to happen to them. Anyway, supporters argued, why would these high-status, upstanding males "need" to rape anyone? They could get all the women they wanted. It wouldn't be until 2015 that Tyson's former manager admitted that such charges were "inevitable" for the boxer, adding that the only surprise was that Tyson was not the subject of further complaints. Several of the Glen Ridge boys eventually went to prison for their crime; another, against whom charges were dropped, joined the military. In 2005 he entered his estranged wife's house, shot her and a fellow soldier, wounding them both, then killed himself—all as his infant daughter lay in the next room. As for Smith? A 2004 assault charge by an employee was dismissed in civil court; in 2005 he settled another suit with a different employee, who accused him of sexual harassment.

Less than a month after Tyson's conviction, the Supreme Court granted students the right to sue colleges and universities for monetary damages under Title IX, which prohibits sex discrimination in education. That gave immediate leverage to young women across the country—at the University of Southern California; Stanford University; University of California at Berkeley; University of Wisconsin; University of Michigan; Tufts; Cornell; Yale; Columbia—who had begun speaking out about campus sexual assault. Most famously, girls at Brown University, frustrated by the administration's indifference, scrawled a list of alleged rapists on the walls of the women's bathroom in the school library. (Boys later retaliated with their own list: "Women Who Need to Be Raped.") Even after the walls were painted black as a deterrent, girls used white paint pens to keep the list going; at one point it swelled to thirty names.

Also during this period, the media began reporting on what was perceived as a sharp and shocking trend of "acquaintance rape" on campus. In December 1990 alone, the *Washington Post* revealed "The Statistic That No One Can Bear to Believe"; *People* ran a cover story on "a crime that too many colleges have ignored"; and Fox TV produced a documentary, *Campus Rape: When No Means No.* As evidence, many pointed to a 1987 study funded by the National Institute of Mental Health and conducted by Mary P. Koss, then a professor of psychology at Kent

State University. Koss surveyed six thousand students at thirty-two universities and found that 27.5 percent—more than one in four—of the girls had, since the age of fourteen, experienced a sexual encounter fitting the legal definition of rape. Eighty-four percent of those attacks were committed by someone the girl knew; 57 percent took place on dates. That led Koss to coin the term *date rape*. When she factored in other forms of unwanted sexual activity ("fondling, kissing, or petting but not intercourse"), the victimization rate shot up to nearly 54 percent. Only a quarter of the boys surveyed admitted involvement in some form of sexual aggression; one in ten said they had verbally pressured a girl into intercourse; 3.3 percent had attempted physical force; and 4.4 percent had raped someone. None of those in the latter two categories considered their acts criminal, largely because they had faced no consequences. "They would say, 'Yes, I held a woman down to have sex with her against her consent,'" Koss told NPR, "'but that was definitely not rape.'" Sexual violence was so pervasive, Koss concluded, that it was part of what the culture defined as "normal" interaction between women and men.

Then came the backlash. In her 1993 polemic, *The Morning After*, Katie Roiphe, a telegenic graduate student in English literature at Princeton, dismissed the campus "rape crisis" as overblown. "If twenty-five percent of my women friends were really being raped, wouldn't I know it?" she reasoned. Perhaps not, considering her main beef: the inclusion in Koss's rape tally of those who answered yes to the question "Have you had sexual intercourse when you didn't want to because a man gave you alcohol or drugs?" To Roiphe, "real" rape involved brute force. Silence alone did not indicate nonconsent; neither did incapacitation. It was a classic conservative argument, and one that endures to this day, but Roiphe gave it a contrarian, "feminist" spin: scolding campus activists for undermining the very agency with which the movement provided them. "A man may give [a woman] drugs," she wrote, "but she herself decides to take them. If we assume that women are not all helpless and naïve, then they should be held responsible for their choice to drink or take drugs." "Rape crisis" feminists, in other words, needed to pull up their Big Girl pants and deal with a few embarrassing nights. Roiphe rejected what she considered their attempt to expand rape's definition as being "a way of interpreting," "a way of seeing" rather than a "physical fact." As if reinterpretation—of citizenship, of suffrage, of who may hold property, even of who are, themselves, property—is not at the core of women's rights: it was just two months before Roiphe's book was published, for instance, that all fifty states finally recognized marital rape as a crime.

Roiphe's book grew out of an editorial in the *New York Times*, which also excerpted it on the cover of its Sunday *Magazine*. Other media outlets (*Newsweek, The Atlantic*, ABC, NBC, PBS) soon began churning out stories and programming on what was suddenly demoted to "the date rape controversy." Few mentioned that even when Koss's data were recalculated without the alcohol question, one in six women had still been legally raped. (To be fair, the statistic was often misstated by activists as the number of girls who would be raped *while on campus*, rather than *since age fourteen*, which is certainly horrifying enough.) When Roiphe lost her novelty, reporters turned to Camille Paglia, who proclaimed, "date rape is bullshit," and Christina Hoff Sommers, currently a resident scholar at the right-wing American Enterprise Institute, whose book *Who Stole Feminism* accused Koss of "[opening] the door wide to regarding as a rape victim anyone who regretted her liaison of the previous night." (Of course, by *excluding* alcohol-facilitated rape, Sommers herself would slam shut the door on "regarding as a rape victim" anyone who was penetrated while passed out drunk.)

By October 1993, campus antirape activism was so maligned that it became fodder for a notorious *Saturday Night Live* sketch, a mock game show called "Is It Date Rape?" Ostensibly set at Antioch College, it lampooned that school's pioneering requirement that partners obtain a clear, verbal "yes" before engaging in sexual activity. Chris Farley, as a frat boy, squared off against Shannen Doherty, as a dowdy "Victimization Studies" major— yes, that's funny—over categories such as "Halter Top," "She Was Drunk," "I Was Drunk," "Kegger," "Off-Campus Kegger," and "Ragin' Kegger." Other cast members, wearing "Date Rape Players" T-shirts, acted out permissible interactions involving such stilted requests as "May I elevate the level of sexual intimacy by feeling your buttocks?" and "I sure had a nice time at that ragin' kegger. May I kiss you on the mouth?" The implication was that the whole date rape thing had gone too far; a bunch of dour, unattractive feminists were trying to shut down the Animal House and ruin heterosexual sex. Days later, the *New York Times*, citing the sketch, weighed in with a staff editorial, scolding Antioch for inappropriately "legislating kisses." Although the director of the school's sexual offense prevention program responded in a letter to the editor that "We are not trying to reduce the romance, passion, or spontaneity of sex; we are trying to reduce the spontaneity of *rape*," the damage was done. "Affirmative consent" (along with Antioch College) became little more than a punch line; date rape was quickly downgraded from "epidemic" to "controversy" to "hype," and further outcry by advocates was essentially quashed. By November of that year, 17.8 million people, mostly teens, tuned in to *Beverly Hills 90210* to see an episode in which Steve, the series' resident goofball, "accidentally" raped a girl who never vocalized the word *no*. She ended up apologizing to *him* in front of a crowd at a Take Back the Night rally. The lesson learned? The "misunderstanding" was actually her own fault because, she said, "I didn't say yes, but I didn't say no, either."

Love and War

Maddie loved Kyle. She did. She'd met him at a party just before her fifteenth birthday; he was a grade ahead of her at a different high school. The two hooked up—nothing serious, just a little kissing. He told her straight out that he liked another girl, though he was willing to be her "friend with benefits." When she saw him at another party a few weeks later, after they'd both been drinking, they hooked up again. Again, they kissed a bit, but this time he informed her he couldn't continue unless she gave him a blow job; otherwise he would develop "blue balls." (Parents, take note: a number of girls I spoke with fell for that chestnut.) Maddie agreed: she had never gone down on a boy, but she already felt she was falling in love with Kyle, and she wanted to make him happy.

Nothing changed between them afterward. Instead, they fell into a pattern: when they saw each other at parties, they would make out, he might "finger" her (though not to orgasm), and she would give him a blow job. They never went out on a date. They never met each other's parents. She'd never even been to his house until they decided to have intercourse. Maddie was sixteen by then and wanted her first time to be with Kyle. They even bought condoms together beforehand, "like a real boyfriend and girlfriend." She remembers the event itself as sweet, but uncomfortable and a little boring: "It hurt really badly for about two minutes, and then I mostly looked at my nails," she told me. She was proud, though, to have had sex sober, during the day, in a bedroom, with someone she loved. Until, that is, a few weeks later, when she heard that Kyle was also having sex with someone else.

Maddie was livid. Her plot for revenge seemed ripped from a script of *Gossip Girl*: that weekend, she would go "looking all hot" to a party where she knew he'd be. And then? She would hook up with one of his friends. "I'm going to *win!*" she remembered thinking. Somehow, though, in a series of convoluted mishaps that don't make sense to anyone over seventeen, she ended up at the wrong place—a houseful of seniors from a neighboring town, mostly boys, none of whom she knew well, all of whom had been drinking. One was a football player named Josh, who had been the sort-of boyfriend of a girl Maddie knew ("like a Kyle-and-me kind of thing"), toward whom he'd been "super abusive." When she informed Josh that the girl had told her "a lot about him," he scoffed. "Don't listen to anything *she* says. She's crazy!" Maddie remained polite but distant, making it clear (at least she thought) that she was uninterested. Somehow, though, word got around the party that the two were going to hook up.

"How did he get *that* message?" Maddie said when another boy asked her about the rumor. "I'm not going to hook up with him. He's an asshole."

The boy smirked. "Well, we'll see after a couple of drinks."

"What does *that* mean?" Maddie replied. "You can't just say that stuff to girls!"

The boy laughed, holding up his hands. "I'm kidding!" he said.

Rape by the Numbers

Throughout the 1990s and early 2000s, research on campus assault quietly continued to accrue, as did skepticism about the results. Using the narrowest definition of rape—as involving physical force—most studies found an annual incidence of between 3 and 5 percent. That is not one in four or even the more recently asserted one in five. Still, given that according to the Census Bureau, there were 4.6 million female full-time undergraduates at four-year institutions in 2013, it would mean that between 138,000 and 230,000 were raped each year—not so comforting. What's more, that conservative definition is no longer employed by such notoriously radical feminist cabals as, say, the FBI, which as of 2013 defined rape as "penetration, no matter how slight, of the vagina or anus with any body part or object, or oral penetration by a sex organ of another person, without the consent of the victim." (That revised definition, incidentally, does not assume the victim is female.)

In 2015 two significant reports came out that should (but probably won't) put an end to all the squabbling. The Association of American Universities' Campus Climate Survey, comprised of over 150,000 students, found that a third of female undergraduate respondents had been victims of nonconsensual sexual contact. Meanwhile, sociologists Jessie Ford and Paula England analyzed assault rates among seniors who had participated in the Online College Social Life Survey. Unlike the AAU report, Ford and England focused solely on acts of intercourse or attempted intercourse—they did not include the incidents of unwanted touching, oral sex, or psychological coercion that critics insist unfairly pad the numbers. Ten percent of the girls said they had been physically forced to have sex since starting college;

15 percent said that someone had tried to physically force them, but that they had escaped without having intercourse (the survey didn't ask whether they had been forced into other acts instead); 11 percent reported someone had unwanted intercourse with them while they were "drunk, passed out, asleep, drugged, or otherwise incapacitated"; and 25 percent reported at least one of these things had happened to them. Including the types of assaults while intoxicated that Roiphe, Paglia, Sommers, and their supporters (if not the criminal justice system) reject, that brings us back to one in four.

Since 1990, colleges and universities have been legally obliged to report to the Department of Education all crimes occurring on or near campus. Those that don't can lose federal financial aid funding, something few schools, no matter how well endowed, can afford. The impetus for that was the rape and murder of nineteen-year-old Jeanne Clery in her Lehigh University dorm room. Clery's parents later learned that there had been multiple violent crimes at the school over the previous three years, but with no consistent tracking policy, students were left oblivious, overestimating their safety on campus. Clery's attacker, who was not a student, had passed through three doors equipped with automatic locks, all of which had been propped open with boxes by dorm residents. Despite that, sanctions for fudging crime stats remained rare, and given that high rates of rape are not a big selling point for prospective students, it's probably not surprising that by 2006, 77 percent of campuses reported their number of sexual assaults at an implausible zero.

That, however, will no longer cut it. In 2011, Russlynn Ali, Obama's new assistant secretary for civil rights, fired off a nineteen-page "Dear Colleague" letter reminding campus officials of their responsibility to uphold all aspects of Title IX, including those involving sexual harassment and violence. Along with a mandate to resolve cases quickly and ensure the physical and psychological safety of accusers (rearranging the class schedule of the accused or removing him from the alleged victim's dorm), the letter laid down a new, reduced burden of proof: "a preponderance of evidence," typically used in civil cases, rather than the more demanding "clear and convincing evidence" then being used on many campuses. More controversy ensued, with conservative activists denouncing the standard as too low given the seriousness of the crime and the potential stigmatization of the accused. The thing is, though, as legal blogger Michael Dorf has written, the lower burden of proof in civil court is not based on either a crime's brutality or its potential to defame the perpetrator, but on the nature of the *punishment*: so someone such as O.J. Simpson could be found not guilty of murder by the standards of criminal court, where life in prison was at stake, but guilty in civil court where the penalty was solely to the pocketbook. Given, then, that colleges expel or suspend rather than jail rapists, "a preponderance of evidence" standard is, in fact, reasonable.

The Department of Education's warning roiled the academic world. As with the right to sue for monetary damages in the 1990s, it also galvanized female students, who no longer needed traditional media to champion their cause: they had the Internet. In 2012, Angie Epifano, a former student at Amherst, published a signed editorial in the school newspaper about college administrators' callous response to her rape allegations. The detailed description of a skeptical sexual assault counselor, her subsequent suicidal depression, a stint in a psych ward, and her ultimate withdrawal from school went viral, generating more than 750,000 page views. "Silence has the rusty taste of shame," she declared. "I will not be quiet." Soon a national movement began to form—activists, often assault survivors themselves, at Amherst, the University of North Carolina, Tufts, Yale, Berkeley—all connecting through social media. That caught the attention of the mainstream press. This round, the *New York Times* seemed all in: running, among other stories, front-page pieces on the student activists and on the White House initiatives; an account in the Sunday Review section by a University of Virginia rape survivor about the lax punishment meted out to her assailant; and numerous opinion pieces and online debates on institutional responsibility, alcohol abuse, the underreporting of assault, and the dubious culture of fraternities and sports teams. The paper also profiled Emma Sulkowicz, a senior at Columbia University who had vowed to lug a fifty-pound dormitory mattress on her back everywhere she went during the 2014/15 school year until the boy she accused of raping her—who had been found "not responsible"—was expelled. (He filed a suit against the university, claiming the administration's failure to protect him from Sulkowicz's accusations, which he said destroyed his college experience and reputation.) Some hailed Sulkowicz as a hero; others called her unhinged. Regardless, it is clear that public witness bearing—rejecting traditional anonymity with its attendant assumption of shame—had become girls' best weapon in the fight against rape.

By the spring of 2015 more than a hundred colleges were under investigation for possible mishandling of sexual assault cases. Among them were the most prestigious in the country: Amherst, Brandeis, Dartmouth, Emerson, Emory, Hampshire, Harvard (the college and the law school), Princeton, Sarah Lawrence, Stanford, Swarthmore, the University of California–Berkeley, the University of Chicago, the University of Michigan–Ann Arbor, the University of North Carolina at Chapel Hill, the University of Southern California, the University of Virginia, and Vanderbilt. Will those inquiries make a difference? It's hard to say. The number of reported campus sexual assaults nearly doubled between 2009 and 2013, from 3,264 to 6,016. Although that wouldn't seem like good news, it is: rather than an increase in the incidence of rapes, the rise appears to reflect a new willingness of victims to step forward, a new belief that they will be heard. The key may be to keep the bright light of public attention shining. According to a study by the American Psychological Association, the reported numbers of assaults increase an average of about 44 percent when campuses are under formal scrutiny. Afterward, though, they sink back to their original levels, indicating that some schools provide a more accurate picture of sexual assault only when forced to do so.

In any case, I would argue that waiting to address rape until college is years too late. Sexual assault is even more common among secondary students; the difference is that their schools don't have the same duty to report it. Twenty-eight percent of female college freshmen in a 2015 survey of a large private university in upstate New York said they had been victims of either attempted or completed forcible or incapacitated rape *before* college—between the ages of fourteen and eighteen. As in the early 1990s, many of the recent incidents that have shocked the nation also took place among younger kids. In the fall of 2012, Steubenville, Ohio, became the Glen Ridge of its day after two football players hauled a drunk, insensible sixteen-year-old girl from party to party, taking turns sexually violating her, spitting on her, even urinating on her as classmates looked on, some cheering. Like the Glen Ridge jocks, who would, without asking their partners, Scotch tape photos to their high school's trophy case of themselves in flagrante delicto, these boys weren't content simply to assault their victim; they needed to document the "achievement." One member of the Steubenville "rape crew" tweeted such gems as "Some people deserve to be peed on" and "You don't sleep through a wang in the butthole" and "Song of the night is definitely Rape Me by Nirvana." Another boy posted a picture of the victim on Instagram, her head lolling back as the boys carried her by her wrists and ankles. In a YouTube video, a laughing young man calls her "deader than," respectively, Nicole Simpson, John F. Kennedy, Trayvon Martin, and the toddler Caylee Anthony. Was online bragging about rape part of a new, ominous trend? A year earlier, a pair of boys in Louisville, Kentucky (fine students and athletes at a prestigious Catholic school), made news when they passed around cell phone pictures of themselves assaulting a sixteen-year-old who lay drunk and semiconscious in her kitchen. Audrie Pott, a fifteen-year-old from Saratoga, California, committed suicide after photos of an assault perpetrated while she was passed out drunk were posted on the Internet. Ditto Rehtaeh Parsons, a seventeen-year-old girl from Nova Scotia, Canada, who was gang-raped while incapacitated.

Tracking those incidents, it struck me how often the words *funny* or, more commonly, *hilarious* came up among boys recounting stories of women's sexual degradation. When, during the Steubenville video an off-camera voice says rape isn't funny, Michael Nodianos, then a high school baseball player, responds, "It isn't funny. It's *hilarious*!" One of the Louisville boys told police he thought it would be "funny" to take pictures of himself assaulting his victim. A young woman I met at a California university told me how, freshman year, a male resident of her dorm invited her to watch a video he'd shot on his phone of a friend having sex with a girl who was out cold. "Come look at this," he had said. "It's *hilarious*." A boy on a midwestern campus I visited, recalling the first time he saw hard-core porn, remembered thinking that was "hilarious," too; his classmate used the word while describing how the "ugly band girls" were the most sexually active in his high school. "Hilarious" seemed to be the default position for some boys—something like "awkward" for girls—when they were unsure of how to respond, particularly to something that was both sexually explicit and dehumanizing, something that perhaps actually upset them, offended them, unnerved them, repulsed them, confused them, or defied their ethics. "Hilarious" offered distance, allowing them to look without feeling, to subvert a more compassionate response that might be read as weak, overly sensitive, and unmasculine. "Hilarious" is particularly disturbing as a safe haven for bystanders—if assault is "hilarious," they don't have to take it seriously, they don't have to respond: there is no problem.

The photos shared by the assailants in Steubenville, Louisville, Nova Scotia, and Saratoga revictimized the girls—potentially in perpetuity, as the images could be endlessly copied, downloaded, and passed along. They also provided unique evidence that crimes had indeed been committed, though that made neither conviction inevitable nor punishments necessarily more severe. One of the Steubenville rapists was given a year in juvenile detention; the other got two years, including credit for time already served. The Louisville boys were ordered to perform fifty hours of community service, which, until the local newspaper intervened, they were fulfilling by putting away equipment after lacrosse practice. Two of Audrie Pott's assailants received thirty-day sentences in juvenile detention, to be served on weekends; a third served forty-five consecutive days. Rehtaeh Parsons's attackers were placed on probation. As in Glen Ridge, there was often a groundswell of sympathy for the boys in these cases: claims that their actions were unusual, a one-time mistake; anguish over the damage convictions would do to their bright futures; denunciations of the girl involved. One of the Louisville assailants took his appeal straight to his victim, texting her to ask that she stop pursuing her case against him. "There is another way to deal with this other than jeopardizing our lives forever. . . . I'm not a bad person just a dumb one."

"You don't think you ruined my life forever?" she shot back.

Uncool

Through another series of jumbled, and by now somewhat alcohol-tinged, events, Maddie found herself in the backseat of a car with Josh, heading to the party she was supposed to have gone to in the first place. A boy named Anthony, another senior, was driving; his girlfriend, Paige, rode shotgun. Maddie ignored them all, focusing on the texts she'd begun trading with Kyle, occasionally yelling at her phone. Josh, seeming truly concerned, asked what was wrong. "There's this guy I've been in love with for a year and a half," Maddie told him, her voice teary, "and I lost my virginity to him and now he's had sex with another girl."

"Have *you* had sex with anyone else?" Anthony asked from the front seat.

"No," Maddie said, still weepy. "I only have sex with people I'm in love with."

"Well, that's your problem!" Anthony told her. "If you just have sex with someone else, you'll get over it!"

Maddie may have been upset with Kyle, but she wasn't stupid; she ignored Anthony's "advice." The group drove around for a while, but they couldn't find the party. Maybe, Anthony said, it had already been broken up by the cops. The boys suggested they head to a familiar park instead, and the girls agreed. When Anthony drove to a wooded area that Maddie didn't recognize, she didn't say anything—she didn't want to appear uncool in front of older kids—but she surreptitiously took a screenshot with her phone. The next day, the Geotag showed the boys had lied: they were nowhere near where they claimed to be. Anthony and Paige strolled off into the trees, leaving Maddie alone with Josh. He pushed her against the car door and began to kiss her. Maddie didn't want to be there, didn't want to be kissing him. She felt angry, confused, maybe a little scared. "God damn it," she thought, "what am I supposed to do?" She tried to tell herself it would be over soon: Anthony and Paige would come back and they'd go to a real party, where she could ditch Josh. When she described what she actually said to him, though, she used the tiny, helpless voice teenage girls lapse into when they're uncomfortable, when they don't want to offend. "I was like, 'Okay, we don't have to keep doing this, get off now!'" she said.

Josh grabbed her wrist and pulled her deeper into the woods. He backed her against a tree and began kissing her again. "I knew this was not good, that I needed to leave," Maddie said. But where could she go? When Josh began pushing down on her shoulders, she shrugged him off. He persisted. She moved his hands. After a few more tries, he finally said, "Oh, is that too hard for you? Do you not want to do it?"

"No, it's not 'too hard,'" Maddie replied, "I just don't want to do *anything* with you." To spare his feelings, she said it was out of respect for her friend, the one he'd called "crazy."

"She doesn't have to know," Josh said.

Maddie shook her head. "No. I just don't want to." At that, she said, Josh began to pout, acting injured and rejected. Just then, Anthony began honking his car horn.

They scrambled into the backseat and Josh brandished a bottle of rum. "I don't know where the top is," he said, "and we can't drive with an open bottle in the car." He thrust it toward Maddie, adding, "So you have to drink it."

Maddie shook her head. She did not want to drink any more.

"Oh, it's okay," Anthony explained. "Rum makes your blood alcohol go up, but you won't feel drunk."

Maybe she was just trying to get through a dicey evening, trying to avoid antagonizing two large, older boys; besides, Anthony wouldn't start the car until that bottle was empty. "Be cool," she told herself as they passed it around. "Just get yourself home, get to bed." She tried faking a few sips, but in the end, she guesses she downed about six shots. After that, her memory fragmented. She recalled crying more over Kyle's betrayal. She remembered going to a fast-food drive-thru. She remembered Josh pulling her onto his lap. And then she blacked out.

Don't Tell Girls Not to Drink; Tell Rapists Not to Rape

At the heart of the argument over consent is another argument over alcohol. How drunk is too drunk to mean yes? How drunk is too drunk to be unable to say no? Who bears responsibility for making that call? An estimated 80 percent of campus assaults involve alcohol, typically consumed voluntarily; often both victim and assailant (or assailants) have been drinking. As I wrote earlier, the party culture on college campuses (as well as in many high school communities) can act as cover for rapists, especially repeat rapists. Yet in 2013, when Emily Yoffe wrote on *Slate DoubleX* that girls should be warned that heavy drinking increases their vulnerability to having sexual violence perpetrated against them, she was pilloried for victim-blaming. *The Atlantic, New York Magazine, Jezebel, Salon, Huffington Post,* the *Daily Mail, Feministing,* and even colleagues at *Slate DoubleX* itself labeled her a "rape apologist." During the ensuing furor, a generation gap emerged. Older women—that is, women the same age as Yoffe (a category that includes this author)—thought her advice sounded sensible. She wasn't, after all, saying that a drunk girl *deserved* to be raped or that it was her fault if she was. Nor was she saying that sobriety guaranteed protection against sexual assault. She only seemed to be voicing what most of us would tell our daughters: alcohol reduces your ability to recognize and escape a dangerous situation. Women metabolize liquor differently from men, too, reaching a higher blood alcohol level drink for drink and becoming more impaired than a guy the same size and weight. Given the prevalence of binge drinking on campus, shouldn't they know that?

Many young women, though, countered with a stance similar to the one they held on dress codes: don't tell us not to drink, tell rapists not to rape. If you really want to reduce assault, they said, wouldn't it be equally, if not more, logical to target *boys'* alcohol abuse, especially since perpetrators are about as likely to be drinking as victims? Alcohol has proven to have a profound influence on would-be rapists' behavior. It lowers their inhibition; it allows them to disregard social cues or a partner's hesitation; it gives them the nerve they may not otherwise have to use force; and it offers a ready justification for misconduct. The more that potential rapists drink, the more aggressive they are during an assault, and the less aware of their victims' distress. By contrast, sober guys not only are less sexually coercive but will more readily step up if they believe an alcohol-related assault is in the offing.

Activists are correct in saying that the only thing that 100 percent of rapes have in common is a rapist. You can shroud women from head to toe, forbid them alcohol, imprison them in their homes—and there will still be rape. Plus, you will live in Afghanistan. To me, this seems like another of those both/and situations. I have a hard time defending *anyone's* inalienable right to get shit-faced, male or female, especially when they're underage. What's that, you say? Harmless collegiate rite of passage? Six hundred thousand students ages eighteen to twenty-four are unintentionally injured each year while under the influence; 1,825 die. Teens who drink in high school, confident in their heightened alcohol tolerance, are at particular risk of harm in college.

I happen to live in Berkeley, California, the town where my state's best and brightest come for their education—the average high school grade point average of incoming freshmen here is 4.46. Yet, in the first two months of the 2013/14 school year, paramedics transported 107 of these smarty-pants students, all perilously intoxicated, to the hospital. During "move-in weekend" alone, the volume of calls about alcohol poisoning to 911 was so high that the city had to request ambulances from neighboring towns; the local ER was overrun with drunk students, forcing diversion of those vehicles elsewhere. (Heaven help the "townie" who happened to have a stroke or a heart attack on one of those nights.) In that same two-month period, incidentally, campus police cited exactly two kids for underage drinking. And yet when binge drinking rises, so does sexual assault. As part of an investigative story by the local ABC-TV affiliate, a paramedic who responded to some UC Berkeley calls, his face blurred and voice distorted to avoid reprisals, told a reporter that he had personally stopped a group of these top-tier college boys as they dragged an unconscious girl out of a party; one admitted he didn't even know her. "Who knows what their intentions were?" the paramedic mused. Nine rapes were reported in the first three months of the 2014/15 school year; five on one night when members of a non-recognized fraternity allegedly slipped "roofies" into their female classmates' drinks, rendering them defenseless.

As a parent, I am all for harm reduction. So I will absolutely explain to my daughter the particular effects of alcohol on the female body. I will explain how predators leverage that difference by using liquor itself as a date rape drug, and how bingeing increases everyone's vulnerability to a variety of health and safety concerns. I know that getting loaded can seem an easy way to reduce social anxiety, help you feel like you fit in, quiet the nagging voice in your head of paralytic self-doubt. Still, knocking back six shots in an hour in order to have fun—or, for that matter, to prove *you* are fun—is, perhaps, overkill. Nor is it ideal to gin up courage to have sex that would otherwise feel too "awkward"—even if the results are consensual, the sex will probably suck. Two people who are lit may *both* behave in a manner they will later regret—or not fully remember, making consent difficult to determine. Should that constitute assault? Students themselves are divided. Nearly everyone in a 2015 *Washington Post*/Kaiser Family Foundation poll of current and former college students agreed that sex with someone who is incapacitated or passed out is rape (a huge and welcome cultural shift). But if *both* people are incapacitated? Only about one in five agree; roughly the same percentage say that is *not* assault, and nearly 60 percent are unsure. That's understandable, given the paradox of students' sexual lives: drunkenness is obligatory for hookups, yet liquor negates consent. There are bright lines—lots of them—and they are too often crossed. But there are also situations that are confusing and complicated for everyone. Recall Holly, who mixed Red Bull and shots (a combination that makes a person appear deceptively sober) before blacking out? Maybe she seemed coherent and eager to have sex; maybe her partner was equally drunk and oblivious; maybe he was stonecold sober and consciously targeted her; she'll never know.

So I'll tell my daughter that it's possible to make mistakes, that not all scenarios are as clear as we would like. That said, if, for whatever reason, she does get wasted—because it's part of the culture she's in or because she wants to see what it feels like or because the drink didn't taste strong—and, God forbid, is targeted for assault, it is positively, in no way, under any circumstances, her fault. I will tell her that nothing ever, ever, *ever* justifies rape. Victims are *never* responsible for an assailant's actions and need not feel shame or be silenced. If I had a son? I would be equally clear with him: drunk girls are not "easy pickings"; their poor choices are not your free pass to sex. I would tell him that heavy drinking, in addition to potential long-term physical harm, impairs boys' ability to detect or respect nonconsent. I would say that if there is *any* doubt about a girl's capacity to say yes—if the thought even flits across his mind—he should, for his own safety as well as hers, move along. There will be other opportunities to have sex (truly, there will be). So although I get why, for both parents and policy makers, focusing on girls' drinking is tempting, it is simply not enough.

"Maddie, You Were Raped"

Later, Paige filled Maddie in on what had happened. The boys dared her to kiss Paige, which she did. Then she kissed Josh, crowing, "I'm the queen of the car because all of you like me the best!"

"If you really want to be queen of the car," Anthony told her, "you have to have sex in the car."

"Okay," Maddie replied, turning to Josh. "Let's do it!"

Maddie insisted on a condom, which made Paige believe the girl was lucid. Anthony, who had one, passed it back to Josh. Maddie remembered, sort of, telling Josh to take off her pants because, drifting in and out of consciousness, she was too drunk to do it herself. She remembered waking up at one point as the car sped through the side streets of her town; she was on top of someone but didn't know who it was or how she'd gotten there. When she realized the person was having intercourse with her, she began to cry. "But I couldn't talk and I couldn't really move," she said. "And I don't think he realized I was crying because he was so into what he was doing." There are more shards of memory, but they are much the same: confusion, tears, incapacitation. Finally, Josh finished, and Maddie rolled to the corner of the car, managing somehow to pull on her pants.

"I want to go home," she said, but the other three were looking for another party.

"No!" Maddie said. "Take me home!"

"What's your problem?" Paige asked, annoyed. "Why are you crying?"

Maddie only cried harder, repeating that she wanted to go home. At that, the other three grew nervous. "Get her out of here," someone said, and they dropped her off, alone, at a strip mall near her house.

The next morning, during an early shift at her job at a neighborhood café, Maddie would periodically start to cry, though she couldn't quite say why. "I knew something bad had happened," she told me, "but I couldn't put my finger on why I was so upset about it." When she got off work, she asked a friend to meet her, and confided what she remembered.

"Maddie," the girl said, "you were raped."

Maddie denied it, but her friend knew Anthony, the boy driving the car, and called him on the spot. "You let this girl get raped in the back of your car!" she told him. He denied it, too, asking to talk to Maddie directly. She remembered his voice as gentle, soothing. "Look," he said, "I know you had a bad night, and you're upset, but you didn't get raped. Stop telling people that."

"I'm *not* telling people that," she said and hung up. When she and her friend got to her home, her friend said she was going to tell Maddie's mom. "I'm sorry," she said, "but I don't know what to do, and someone needs to take care of this."

Maddie went to her room so she wouldn't have to see her parents' reaction. A little while later, her father knocked on the door, a notebook in hand. She told him the story in as much detail as she could muster.

"Why didn't you say no?" he asked.

"I did!" she said. "But then I just got drunker and . . . I don't know. I can't explain." Maddie didn't go back to school that Monday, or the day after that, or the day after that. She hardly got out of bed for a week. Paige, meanwhile, began spreading rumors, claiming that Maddie had cried rape because she was embarrassed to have lost her virginity in the back of a moving car. Strangers on Facebook posted that Maddie was "a lying whore." Few classmates, boys or girls, took her side. "None of them knew what actually happened," Maddie said. "*I* didn't know what actually happened. I still don't. There are still parts of the story I'm not clear on."

Not even her (now former) friends stood by her. "They'd say, 'I wasn't there, so I can't judge if it was true or not.' And I'd be like, 'Why aren't you just taking my side? I thought we were friends!'" Josh, unsurprisingly, called her a liar, too. He did contact her directly once, via text, early on. "Are you telling people I raped you?" he asked. She texted back that she was not. He never got in touch again. "Obviously no guy is going to admit to that," she said. "I don't expect him to. I don't expect him to ever apologize. Why would he? In his eyes he didn't do anything wrong. It's not like he took me to a dark alley to rape me. He just really wanted to have sex, and I said no, and it hurt his pride."

One of the only people to stick by Maddie was Josh's former girlfriend—or hookup buddy, or whatever she was: the one Maddie said he'd treated badly. "She believed me without question," Maddie said. "That stuff with him pushing on my shoulders? He did that kind of thing to her, too. And there have been two other girls who told me he's done similar stuff to them. But mine was the only time it turned into this huge mess." Maddie shook her head and sighed. "I think he'll get in trouble at some point, though."

Christmas break came, and Maddie hoped that, with it, the incident would be forgotten; it wasn't. As December turned to January and classes resumed, the gossip spiraled out of control: Maddie was pregnant! Maddie had had an abortion! She withdrew from school and stopped going online or checking texts. Eventually, she enrolled here, at the community college. At least one of her female classmates, she has discovered, was there for the same reason.

What Yes Means

One of the Big Bads that conservatives warned of in the 1990s was that if alcohol-induced assaults were included in the definition of rape, college administrators would be swamped by vengeful girls who regretted their previous night's encounters. As if it's easy for a victim of sexual assault to come forward. As if girls have been readily believed. As if it weren't social suicide. As if they wouldn't be shunned, called sluts, blamed, harassed, and threatened. Consider the reaction in 2014 on CollegiateACB, a forum where students anonymously discuss campus issues, after a Vanderbilt University student's rape accusations resulted in the suspension of a fraternity. Forum users demanded to know the identity of "the girl who ratted"—a name was actually posted—and called her, among other things, "manic depressive," "a crazy bitch," "psycho," "NASTY AS SHIT," "a no good CUNT," and, over and over, a "snitch." "This repeated use of the word 'snitching' in the thread," wrote André Rouillard, editor of the school's newspaper, "implies that the victim has revealed a secret that should have been kept hidden behind closed doors—under the rug and on floors that stick like flypaper and stink of old beer. . . . The OP [original poster] issues a rallying cry: 'we need to stick together and prevent shit like this from being ok.'" By "shit like this" he didn't mean rape; he meant girls' reporting of it.

Those trying to prove that campuses are rife with psycho young women just itching to ruin their male classmates' lives were inadvertently handed an opportunity in the spring of 2015, when *Rolling Stone* magazine retracted an article on a gang rape at the University of Virginia that had fallen apart under scrutiny. I don't know if that scandal will become the cornerstone of a new suppression of activism—these are different times than the 1990s—but as a Columbia University Graduate School of Journalism investigation concluded, *Rolling Stone*'s editors "hoped their investigation would sound an alarm about campus sexual assault and would challenge Virginia and other universities to do better. Instead, the magazine's failure may have spread the idea that many women invent rape allegations."

There are, absolutely, false charges of rape. To say otherwise would be absurd. But they are rarer than alarmists would like you to believe. Legally, a "false report" is one in which it can be *demonstrably* proven that a rape was not committed. When investigators find that assault did not occur, that is something else: an unsubstantiated or inconclusive report. Conservative pundits such as Hoff Sommers, Cathy Young, and Wendy McElroy—plus every troll ever on the Internet—assert that 40 to 50 percent of sexual assault accusations are actually *false*. (Although, oddly, as criminologist Jan Jordan has pointed out, while adamant that half of accusers lie, such critics believe women who recant are unfailingly truthful.) In her book *Rape Is Rape*, Jody Raphael explains that this statistic comes from a 1994 report for which Eugene J. Kanin, a sociologist at Purdue University, compiled one police agency's characterizations of forty-five assault claims made over nine years in a small midwestern town—assessments that were not necessarily based on evidence or investigation. Kanin himself cautioned that his findings should not be generalized, and admitted, "Rape recantations could be the result of the complainants' desire to avoid a 'second assault' at the hands of the police." More credible, Raphael wrote, are seven rigorous studies conducted in the United States and the United Kingdom over more than three decades. They place false claim rates at between 2 and 8 percent, a number, according to FBI statistics, that has been steadily dropping since 1990, when the controversy over acquaintance rape emerged. Certainly it is important to bear in mind the potential for false claims, but our fear of them seems strangely disproportionate, especially given that most victims are not believed, that 80 percent of campus rapes are never even reported, and a mere 13 to 30 percent of assailants are found responsible among the sliver that are.

Emily Yoffe, who also raises the specter of an "overcorrection" on campus rape, has objected that lumping psychologically coerced or pressured sex into statistics risks "trivializing" assault. She, too, fears it would tempt any girl who "regrets making out with a boy who has 'persuaded' her" to file a complaint that could lead to his expulsion. "We may be teaching a generation of young men that pressuring a woman into sexual activity is never a good idea," she acknowledged, "but we are also teaching a generation of young women that they are malleable, weak, 'overwhelmed,' and helpless in the face of male persuasion."

This is where she and I part ways. Most sexual interludes among high school or college students are, obviously, not violent: They are consensual and wanted, if not always reciprocal. That said, a sizable percentage is coerced; rather than "trivializing" rape, Yoffe risks "trivializing" the way such pressure is seen as a masculine right and how that shapes our understanding of consent— even of sex itself. Despite changing roles in other realms, boys continue to be seen as the proper initiators of sexual contact. (If you don't believe me, listen to the outrage of mothers of teen boys when discussing today's "aggressive" girls.) Boys' sex drive is considered natural, and their pleasure a given. They are supposed to be sexually confident, secure, and knowledgeable. Young women, as I've said, remain the gatekeepers of sex, the inertia that stops the velocity of the male libido. Those dynamics create a haven for below-the-radar offenses that make a certain level of sexual manipulation, even violence, normal and acceptable. I don't know that such acts deserve expulsion, but they are worthy of serious discussion. As Lorelei Simpson Rowe, a clinical psychologist at

Southern Methodist University who works with girls on refusal skills, explains, "The vast majority of sexual violence and coercion occurs in situations that are not obviously dangerous . . . so if nine times you go out with a boy and engage in consensual activity, and it's pleasant and you're excited to be developing a relationship, that doesn't prepare you for that one time when it switches."

While such transformations may be sudden, frequently, Simpson Rowe says, they're not. "Guys will start saying, 'Come on, let's go further' or 'Why not?' or 'I really like you. Don't you like me?' There's a lot of persuading and pleading and guilt-inducing tactics, along with a lot of complimenting and flattery. And because it's subtle, you see a lot of self-questioning among girls. They wonder, 'Am I reading this right?' 'Did he actually say that?' 'Did he actually mean that?'" Simpson Rowe and her colleagues have developed a training program that uses virtual reality simulations to help girls recognize and resist those cues. In pilot trials of high school and college students, incoming participants generally rated themselves as confident that they could rebuff unwanted advances or escape threatening situations. Yet, when role-playing a range of increasingly fraught scenarios—from a male avatar who badgers girls for their phone numbers to one who threatens violence if they don't submit to sex—they would freeze. Simpson Rowe was quick to say that only perpetrators are responsible for assault, but assertiveness and self-advocacy are crucial defensive skills. "What we found is the importance of women being able to make quick, cognitive switches between normal sexual interaction and protecting their safety," she said. "And part of that involves being able to notice when something has gone from being a normal interaction to pressure."

The girls in her program worried that a direct rejection would hurt boys' feelings; they felt guilty and uncomfortable saying no.

"Girls have all this modeling for being nice and polite and caring and compassionate about others' feelings," Simpson Rowe explained. "These are wonderful things—good characteristics. But because they're so ingrained, a lot of women think this is how they're supposed to be when faced with an unsafe situation, and they're afraid of being seen as rude. The word that comes up a lot is *bitchy*. So, it's kind of an 'aha' moment when they realize a guy who is pressuring and persuading and not stopping when you say you don't want to do something is not respecting you or your boundaries—and at that point, *you don't have to worry about hurting his feelings.* We emphasize how early the coercive process begins and help them respond to it before it ever gets to violence." Preliminary data showed that three months after completing the ninety-minute training, participants had experienced half the rate of sexual victimization than a control group. Another risk-reduction program piloted among more than four hundred fifty Canadian college freshmen had similar results: a year later, rates of rape among participants were half that of girls who had only received a brochure. "We want to send the message that no one has the right to push or pressure you into what you don't want to do," Simpson Rowe said. "You have the right to stand up for yourself as loudly and physically as you want to and can."

Listening to Simpson Rowe, I thought about Megan, who told her rapist, "Thanks, I had fun." I thought about another girl I met, a freshman in college, who told me her high school boyfriend had raped her twice—once while they were together and once after they'd broken up, when he lured her into his car at a party to talk. Both times, she was drunk. Both times she told him no. Both times he ignored her. "I probably could have pushed him off of me or rolled over or screamed loud enough so someone could hear," she said, "but something prevented me from doing it each time. I'm a very strong person. I have very strong morals. I'm not embarrassed about talking about anything. But I didn't do anything. It was kind of like being paralyzed." I recalled Simpson Rowe's words again in the summer of 2015, when I read the court testimony of a former student at St. Paul's prep school in New Hampshire. A popular senior boy had assaulted her in the spring of her freshman year, she recounted, during an end-of-year rite known as "the senior salute," in which graduating male students compete to have sexual encounters with as many younger female students as possible. Initially flattered by his attentions, she testified, she joined him in a dark maintenance room but was at a loss as to how to respond to his escalating aggression. "I said, 'No, no, no! Keep it up here,'" she told the jury, gesturing to the area above her waist. "I tried to be as polite as possible." Even as he groped, bit, and penetrated her, she said, "I wanted to not cause a conflict."

Each of those girls could have used a session in Simpson Rowe's virtual reality simulator. At the same time, I also thought about a 2014 study in which nearly a third of college men agreed they would rape a woman if they could get away with it—though that percentage dropped to 13.6 percent when the word *rape* (as opposed to "force a woman to have sexual intercourse") was actually used in the question. Teaching girls to self-advocate, to name and express their feelings in relationships, is important for all kinds of reasons, and it may indeed help some of them stop or escape an assault. Yet, just as focusing on girls' drinking disregards rapists' behavior, keeping the onus on victims to repel boys' advances leaves the prerogative to pressure in place; it also maintains sexual availability as a girl's default position even if, as feminist pundit Katha Pollitt has written, she "lies there like lox with tears running down her cheeks, too frozen or frightened or trapped by lifelong habits of demureness to utter the magic word." Even if that girl were to say no loud and clear, the boy might not hear it.

"Affirmative consent" policies—versions of the one pioneered by Antioch—have once again become the hope for change. In 2014, California was the first state to pass a "yes means yes" law directed at colleges and universities receiving state funds. Rather than requiring an accuser to prove she said no, it demands that an alleged assailant prove that there was "an affirmative, unambiguous, and conscious decision by each participant to engage in mutually agreed-upon sexual activity." In other words, that a clear, enthusiastic "you bet," either verbally or through body language, was given. Consent may also be revoked anytime, and a person incapacitated due to drugs or alcohol is not legally able to give it. That's a fundamental shift in power relations, and twelve years after the "Is It Date Rape?" *SNL* sketch, fewer people are laughing. New York passed affirmative consent legislation in 2015. New Hampshire, Maryland, and Colorado are all considering similar bills. Every Ivy League school except Harvard now has a version of "yes means yes" in place as well.

Conservatives have predictably warned that thousands of boys will soon be ejected from colleges for trying for a good-night kiss. But the policies have made liberals uneasy as well. Ezra Klein, editor in chief of Vox, wrote that he supported the law, though he believed it would "settle like a cold winter on college campuses, throwing everyday sexual practice into doubt and creating a haze of fear and confusion over what counts as consent." The anxiety on both sides reminded me of the 1993 fears about California's then-innovative law against peer-to-peer sexual harassment in schools, which allowed districts to expel offenders as young as nine years old. But you know what? Twenty-plus years later, no fourth-graders have been shipped off to San Quentin for hazarding a playground smooch. Nor have school districts been bankrupted by a deluge of frivolous lawsuits. At the same time, the legislation has not stopped sexual harassment. It has, however, provided a framework through which students can understand and discuss the issue, and the potential for recourse, on a number of levels, when it happens. Remember Camila Ortiz, the girl who called out her vice principal when he told girls to cover up and "respect yourself"? She and a friend later organized a group of girls *and* boys to fight sexual harassment at their school. In the winter of 2015, group members addressed a meeting of the school board, presenting a petition signed by more than 750 students, both female and male; among their concerns was that the high school was out of compliance with both state and federal laws. The district's policy is now being redrafted. No one was expelled; no one was sued; no one went to jail. Plus, the students got a great lesson in civic responsibility, leadership, and making social change. Increased awareness has also reduced tolerance for the winking acceptance of harassment and assault. Anheuser-Busch found that out in 2015, when the company unveiled a new tag line for Bud Light: "The perfect beer for removing 'no' from your vocabulary for the night." American sensibilities had changed since the 1990s, as had the targets of influential comedians' humor. So, rather than mocking overly sensitive women, John Oliver drew cheers from his college-age studio audience by skewering the frat-boy mentality that allowed the slogan's approval: imagining Bud executives fist-pumping and shouting, "Sick idea, brah!" "That's what I'm talkin' about, *a'ight*," "*No, no, no, no.* That's what *I'm* talkin' about, son!" and a wordless, "*Blaaaaaaaaaaaaah!*" (The beer company had been forced several days prior to issue a public apology after news of the slogan had careened around Twitter.)

Will affirmative consent laws reduce campus assault? Will cases be more readily resolved? I can't say. As Pollitt pointed out, adjudication in many instances will still be based on he said/she said, with accused assailants replacing "She didn't say no" with "Dude, she said yes!" Among the students in the *Washington Post*/Kaiser Family Foundation poll, only 20 percent said the yes means yes standard was "very realistic" in practice, though an additional 49 percent considered it "somewhat realistic." What "yes means yes" may do, though, especially if states aim solid curricular efforts at younger students, as California plans to, is create a desperately needed reframing of the public conversation away from the negative—away from viewing boys as exclusively aggressive and girls as exclusively vulnerable, away from the embattled and the acrimonious—and toward what healthy, consensual, mutual encounters between young people ought to look like. Maybe it will allow girls to consider what they want—what they *really* want—sexually, and at last give them license to communicate it; maybe it will allow boys to more readily listen.

THAT WAS THE hope of a Bay Area nonprofit that invited me to observe a focus group of high schoolers convened on a November afternoon to discuss consent.

The kids—two African American boys, two white boys, two white girls, a Latina, and an Asian girl—sprawled across couches in a borrowed living room, their conversation subtly guided by a twenty-something facilitator. Over the course of several hours, they wrestled with how alcohol-fueled hookups made "yes" feel like a moving target; with the social costs of saying a direct "no"; with the awkwardness of intervening when a drunk friend was hurtling toward regret; with how they negotiated, or didn't, consent in their long-term relationships. They talked about assault, too. Two of the girls had experienced some form of violation; another was trying to come to terms with troubling accusations by one close friend against another. One of the boys, too, had been lured into sex by an older classmate when he was too drunk to refuse. He wanted to know: was that rape?

More often, though, they talked about the complexity of establishing basic boundaries, with partners and within themselves, in a culture of contradiction, in which there has been some, but not enough, change in the expectations for, consequences of, and meaning of sex for both boys and girls. "Like, okay, 'yes means yes,'" said Michael, who had pushed his shaggy hair, Mark Sanchez style, back in a headband. "But how does that 'yes' change with every situation you're in? When you're drunk, what does that 'yes' mean? Or is it only really 'yes' when you're sober?"

"And what about people getting drunk in *order* to say yes?" Annika added, sitting forward eagerly, her elbows resting on her knees. "I know a situation where two people were interested in each other and asked a friend to have a party so that they could get drunk and hook up."

Caleb, who had a "fade" haircut and red plastic glasses, jumped in. "The whole problem is that hooking up sober is not so attractive."

Annika nodded and continued. "And yes can mean different things, *especially* if I'm drunk. Like, did I say yes because I wanted to hook up with *this* person or because I wanted to hook up with *someone*, or because my friends think it would be cool if I hooked up with that person?"

Nicole confided that when her "gut" told her to end a hookup, she would immediately start a mental tally of everything she had done up until that point—locked eyes with a boy across a room, flirted, touched his shoulder, kissed him, taken off her shirt—that would have led him to believe she would say yes to more. "And I'm already feeling guilty and worrying about what will happen in that moment of confrontation when I actually say to him, 'This is my boundary.'"

"It's so complex," said Gabriel, who wore a five-panel cap and a U.S. Marine Corps T-shirt. "As a guy, you have to do the best *you* can do to prevent a situation from happening in the future. You have to train yourself to look at someone and say, 'Are you okay with this? Are you one hundred percent sure? Is this definitely a yes?'"

Lauren, who had recently broken up with her boyfriend, quietly offered that even in a long-term relationship, consent could feel tricky. "It's like if you've had sex once, you've said yes forever," she said, and two other girls nodded. "And it's always going to end that way no matter what was voiced or what was wanted at that moment, because once you get to that point with someone, that's what always happens." "Good girlfriends" say yes, no matter what. They consent—or at least comply—freely, even if the sex is unwanted. They take one for the team to keep their relationships stable, their partners happy. What, these young people wondered, do you call that?

"You know," Michael said, "hearing all this . . . I was in a relationship for about a year and I think . . . I was probably on the other side of that equation. I think . . . I didn't mean to, but I was probably subconsciously pressuring my girlfriend." He fell silent for a moment, pondering that. "I don't know that I want to be, like, a leader in gender equality," he continued, "but whatever I end up doing, wherever I end up going, this is going to be something I incorporate. I think just by doing that with the people you meet at school or the people you work with that you can have considerable influence in changing a culture, a community. I really do."

"I Know What It Feels Like to Be Told, 'It's Not Rape.'"

"Do you think you were raped?" I asked Maddie.

She gazed down at her fingers and shrugged. I considered the decades of argument behind that question: not long ago, the answer, maybe my own answer, would have been a definitive no. So much had changed, and so much had not. "Legally?" Maddie asked. "Yes, I was. Asking for a condom doesn't imply consent. But the way everyone treated me afterward . . ." She shrugged. "People will say, 'You had to switch schools because of *that*? That's *nothing*.' And guys are like, 'Oh, that's not rape.' So, I don't know." Maddie fell silent a moment. "Lately, I've been writing blog posts and articles on changing 'rape culture.' Because I know what it feels like to be told, 'It's not rape.' And I know how horrible it was afterward. If I can prevent that, or worse, from happening to someone else, that's all I want to do."

Maddie had been careful during our conversation never to use her assailant's real name. At one point, though, she slipped, and once I was home, it was the work of a moment to find him online. He'd been on the basketball and track teams in his high school, appeared to be a solid student. He'd joined a frat this year, as a college freshman. None of that meant he'd assault someone, though both his history and interests put him at risk: fraternity brothers and athletes are disproportionately represented among repeat offenders. My eyes fell on the name of the large university he attended. I had, at that time, eight nieces who were also college students. It chilled me to realize that he was in school with one of them.

CHAPTER 7

What If We Told Them the Truth?

Charis Denison stood before seventy tenth-graders in the all-purpose room of a Northern California high school. A blond woman in her early fifties, permanently tanned from a former career as a wilderness ranger, she was barefoot, having kicked off her boho-chic wedge sandals, and was wearing her habitual tunic and jeans. A silver chain encircled one ankle, and a beaded mesh bracelet wound up her left arm. On her right hand, above a stack of jangling bangles, she sported a plush, anatomically correct vulva puppet. At the moment, her finger was fondling its clitoris as she commented, "I talk to so many girls where the first person to actually touch their clitoris is somebody else." There have been times over the past two hours when the students—both boys and girls—who were sprawled across the carpeted floor, were a little squirrelly, a bit inattentive. Now, though, they were rapt. "It's hard when you're trying to have a sexual experience with someone and you don't know what feels good to *you*," Denison said. "It's hard to let someone else have that power to decide. So if someone is choosing to become sexually active with someone else, it's really good to be sexual with oneself first. It's good to figure out what you like."

That's right. Denison just encouraged teenage girls to masturbate, and she did it in front of teenage boys. She told the whole class not only that girls have clitorises but what those organs are for—the *only* thing that they are for: to make them feel good. And that, in the annals of American sex education, is nearly unheard of. Denison doesn't call herself a sex educator, though. She sees herself as a "youth advocate," providing accurate information and a nonjudgmental forum in which kids can discuss sex and substance use along with larger ideas of ethics and social justice. She travels to high school communities across California—most, given her frank approach, are private like this one, though an increasing number are public—visiting each class several times a year, building cumulatively on what came before. Her curriculum incorporates decision making, assertiveness skills, sexual consent, personal responsibility, gender roles, and the diversity of sexual orientation and gender identity. But "my job," as she told today's tenth-graders, "my *whole* job is to help you make as many decisions as possible that end in joy and honor rather than regret, guilt, or shame."

Denison talks about risk and danger in her classes (though she doesn't necessarily use that language). She addresses anatomy and contraception, if those aren't part of students' regular health curriculum. By graduation, even if her students plan to stay abstinent until marriage ("which is *awesome!*") or will never have sex with a man, she expects them, nonetheless, to be able to put on a condom, "drunk, dizzy, and in the dark." She also talks about something usually omitted in the parental "talk" and by the football coaches who, inexplicably, teach "health": sexual activity should be a source of pleasure for teenagers. Not only is hers a more honest perspective, but she believes (and research confirms) that it is ultimately the most effective strategy for reducing risk. "To some parents in school communities, that doesn't sound right," Denison told me, "but it *is* right. [Teens] abstain with more information because they have options, because they have knowledge, because they have alternatives. It's so clear to me that in this area the less specific and the less open we are, the more and more at risk we're putting these kids—especially girls."

Denison's approach is controversial, so controversial that I had a hard time finding a school that would let me observe her in action. Her philosophy doesn't exactly jibe with the just-say-no thinking that's dominated sex ed for the last three decades, but it's one that is slowly, gradually gaining credence. In 2011 the *New York Times Magazine* profiled Al Vernacchio, a revolutionary Philadelphia educator who famously compares sex to eating a pizza: Both start with internal desire—with hunger, with appetite. In both cases, you may decide, for any number of reasons, that it's not the right time to indulge. If you do proceed, there should be some discussion, some negotiation—maybe you like pepperoni and your dining companion doesn't, so you go halfsies, or agree that one person will get his pick next time, or choose a different topping altogether—and a good-faith effort to satisfy everyone involved. There is no rounding bases in that metaphor, no striking out. The emphases are desire, mutual consent, communication, collaboration, process, and shared enjoyment.

Similarly, in 2009 the Population Council published the *It's All One Curriculum*, downloadable for free online, created in conjunction with, among others, the United Nations General Assembly, the World Health Organization, UNAIDS, and UNESCO. Integrating ideas about human rights and gender sensitivity, these guidelines aim to help educators and others "develop the capacity of young people to enjoy—and advocate for their rights to—dignity, equality and responsible satisfying and healthy sexual lives." That curriculum, like Denison's and Vernacchio's, presents sexual exploration (whether alone or with others) as a normal part of adolescence. Sure, there are hazards, but there are also joys, and our role as caring adults is to help our kids balance the two. I admit that, as a mom, the idea of my child becoming sexually active is only marginally less mortifying than the thought of my parents doing anything beyond the three reproductively necessary acts it took to conceive my brothers and me. But the consequences of parental silence, classroom moralizing, and media distortion are far worse. There has to be a better way.

Strange Bedfellows: Sex and Politics

In 1959 abortion was still criminal. Unmarried women could not legally procure contraception, and pharmacists, according to sociologist Kristin Luker, author of *When Sex Goes to School*, would refuse to sell condoms to men they thought were single. Although, even then, over half of women and three-quarters of men had intercourse before their wedding day, there was broad public agreement that sex should be reserved for marriage. That was about to change—radically and quickly. The introduction of the birth control pill in 1960 was the first salvo in the sexual revolution. That was followed three years later by the publication of *The Feminine Mystique*, which launched a new wave of feminism. A decade after that, the Supreme Court guaranteed women's right to abortion. As sex became untethered from reproduction, the notion of "waiting until marriage," or even until adulthood, grew increasingly obsolete: between 1965 and 1980 the percentage of sixteen-year-old girls who'd ever had intercourse doubled. A group of activists, led by Mary Calderone, the physician who founded the Sex Information and Education Council of the United States (SIECUS), hoped those changes would herald an age of positive, value-neutral, medically accurate sex education.

That was not to be. Instead, according to Jeffrey Moran, author of *Teaching Sex*, to ensure minors' ongoing access to contraception, congressional liberals skewed negative, popularizing the idea that teen sex, while perhaps inevitable, was inherently risky and a "crisis" requiring damage control. They argued that the "epidemic" of teen motherhood triggered by the new sexual freedom was, particularly among African Americans, responsible for spiraling poverty. (In truth, although the birth rate among black girls was three times higher than among whites, the overall teen birth rate dropped steadily through the 1960s and 1970s.) The only pragmatic response was to teach kids to protect themselves. So the Adolescent Health Services and Pregnancy Prevention and Care Act of 1978, introduced by Senator Edward Kennedy, while perpetually underfunded, championed educational programs that would focus on risk management, contraception, abortion education, counseling, and "values clarification." It also established a murky, nonspecific idea of "readiness," rather than marriage, as the expected standard for sexual behavior. That, Moran wrote, infuriated conservatives. As Diane Ravitch, an educational consultant and activist, railed (inaccurately, by the way), "Is it appropriate for the government to teach its citizenry how to masturbate? To explain how to perform cunnilingus? To reassure them that infidelity is widespread?"

With that, sex education, previously relegated to innocuous "Family Life" classes, where it was embedded in lessons on successful marriage, became a battleground: a vector for right-wing trepidation about the erosion of traditional matrimony, the rise of women's rights, the growing acceptance of homosexuality, even the potential dismantling of gender itself. In 1981, partly as a reward for the New Right's support of his presidential bid, Ronald Reagan signed what was nicknamed "the chastity law," the first legislation requiring that federally funded sex education, as its sole purpose, teach "the social, psychological and health gains to be realized by abstaining from sexual activity." Reagan, however, allocated only $4 million a year to the bill; it wasn't until the Clinton administration—oh, the irony!—that annual funding for abstinence education shot up to $60 million, a slab of pork tucked into the 1996 Welfare Reform Act. As the money grew, the message it promoted became even more restrictive: to get the cash, public schools would now have to teach that marriage was the *only* acceptable sphere for physical relations, and that sex outside of it at any age (including after divorce or widowhood) would lead to irreparable physical and emotional harm.

Under George W. Bush, the funding for abstinence-until-marriage programs continued to rise, reaching, at its peak, $176 million a year. So it was that in 1988, when the AIDS epidemic was in full swing, only 2 percent of sex ed teachers taught abstinence as the best way to prevent pregnancy or disease, yet by 1999, 40 percent of those supposedly teaching comprehensive sex ed considered it the most important message they were trying to convey. By 2003, 30 percent of public school sex education classes provided *no information whatsoever* about condoms or other contraceptives (beyond their failure rates) and by 2005 over 80 percent of federally funded abstinence-only programs were found by a congressional report to be teaching blatantly inaccurate information, including such "facts" as that the Pill is only 20 percent effective in preventing pregnancy, that latex condoms cause cancer, that HIV can be transmitted through sweat or tears, and that half of homosexual teen boys already have the virus.

All together, the federal government has spent $1.7 billion plus on abstinence-only programs since 1982; that money might just as well have been set on fire. As I mentioned earlier, while virginity pledgers delayed intercourse for a few months longer than their nonpledging peers, when they did become sexually active, they were less likely to protect themselves or their partners against pregnancy or disease. The same holds true for participants in abstinence-only classes. Studies stretching back over a decade have found that, at best, when compared to a control group, participants neither abstain entirely from sex nor delay intercourse; they also do not have fewer sexual partners. They are, however, a lot more likely to become unintentionally pregnant: as much as 60 percent more likely. That could lead one to suspect that abstinence-only advocates are more concerned with ideology than with public health or even sexual restraint— otherwise they would have given it up long ago for something that has been repeatedly proven to reduce teens' sexual activity, increase their use of contraception and disease protection, and improve their relationships: comprehensive sex education.

Under President Barack Obama, comprehensive sex ed finally got its first federal love, although the focus remained squarely on reducing negative consequences: $185 million earmarked for research and programs that have been shown, through rigorous evaluation, to reduce teen pregnancy. That money, of course, could easily disappear under another, less progressive commander in chief, and probably will: for instance, a clause buried in the Student Success Act, a Republican rewrite of No Child Left Behind that passed the House in the summer of 2015, zeros out any funding for programs that "normalize teen sexual activity as an expected behavior, implicitly or explicitly, whether homosexual or heterosexual." Meanwhile, $75 million in abstinence-only funds continued to be doled out each year through the Affordable Care Act. While substantially less than under President Bush, that's still an awful lot to blow on the sex ed equivalent of a tinfoil hat.

What this means for parents is that you never know what your child's "sex education" class may entail. Only fourteen states require that sex ed be medically accurate. Yet even that is no guarantee. Mine is supposed to be one of them. Yet it wasn't until the spring of 2015 that a judge ruled for the first time against a public school system that was actively teaching misinformation: students in the city of Clovis, California, had for years been made to watch videos that compared an unmarried woman who had intercourse to a "dirty shoe" and were encouraged to chant the antigay motto "One man, one woman, one life." Around that same time, Alice Dreger, a professor of medical humanities and bioethics at Northwestern University's Feinberg School of Medicine, live-tweeted her son's abstinence-based sex education class from a public high school in politically progressive East Lansing, Michigan. Instructors there warned about the potential failure rates of contraceptives, citing, as an example, a box of condoms in which *every single one had a hole*! They also advised the boys to seek out "good girls" who say no to sex. At one point, Dreger tweeted, the students were told, "'We are going to roll this dice 8 times. Every time your number comes up, pretend your condom failed and you get a paper baby.'" She followed that up shortly with "Paper babies are being handed out to EVERYONE. They have ALL HAD CONDOM FAILURE AND THE WHOLE CLASS IS PREGNANT." When it was over, Dreger tweeted, "I just want to grab all those kids after school and say HERE IS

THE TRUTH. SEX FEELS GOOD. THAT'S WHY YOU SEEK IT. TAKE CARE & HAVE FUN."

Life Is Like an English Essay

Nearly twenty-five years ago, while teaching English and leading outdoor programs at an all-girls' private school, Charis Denison had an epiphany. So many of the critical lessons of middle and high school took place outside the classroom. Her students wanted (needed) to talk about their experience, but didn't know how, and anyway, there was nowhere they could try. "I started feeling that we were failing these kids," she told me. What would happen if she carved out a formal space for those conversations? What would happen if she encouraged students to apply the rigorous critical skills they used in the classroom to life beyond it? "You wouldn't walk into an essay exam wondering which book the test was on, right?" she said. "But people will go to a party without any thought at all, not even of what they *don't* want to happen." Rather than blaming themselves when things go awry, students needed to remember, Denison began urging them, the "reflect-revise-redraft" strategy they used when editing a paper. "Instead of just thinking, 'Oh my God, that night was awful, that was horrible!' I want them to back it up and think, '*Why* did that suck? And what part did I play in it, and what part was out of my control?' Just like you would with a bad grade, or with anything else that goes wrong. Avoiding the blame game—just backing up, figuring it out, reflecting on it, revising your plan, forgiving yourself, and moving forward."

Denison consciously avoids labels like "good" and "bad," "responsible" and "irresponsible," even "healthy" and "unhealthy," in her classes. "Those are a matter of personal belief," she explained. "The idea of 'regret' works regardless." That's important, she said, because she teaches in communities that encompass a broad range of backgrounds and values. During time devoted to anonymous questions, one student might want to know, "Is it okay that I have casual hookups on a pretty regular basis?" Another, in the same class, might ask, "Is it okay if I wait until I get married to have sex?" "In that context, the idea of 'good' choices doesn't make sense," she explained. "What's key is to be able to talk about sex in a way that makes it equally comfortable for both of those students. So, if Monday morning, after you hooked up with a couple of guys, you feel joy, then that's the right choice. And from there we can back up and ask: Is that serving your partners, too? Is it clear that you're on the same page? And if you're not, does that really serve you? Then, for that girl who perceives sex as something she's holding on to and wants to keep as a part of her to give to a partner that she made a commitment to be with for the rest of her life: How does that feel? If you don't feel guilt, if you don't feel shame, if you're feeling joy and honor, then *bingo*. And if you are feeling guilt and shame, then let's talk about that. Where is it coming from? So developing this idea of 'How are the choices affecting me *and* the people around me? How are they serving me, and how are they serving my partner?'"

Much of Denison's curriculum, perhaps most of it, is not specifically about sex. It's about decision making and communication, skills that are useful in any realm. On another afternoon, I watched her with a group of ninth-graders she was meeting for the first time. She was explaining what she calls "fallbacks," unconscious, reflexive behaviors we resort to when we're uncomfortable. "A lot of fallbacks come from gender roles," she said. "A lot of them come from ways we cope in our families. Like, what if you want to do one thing after school and your friends want to do something else? Or you're in a situation and all of a sudden you're super uncomfortable and you don't know what to do?"

Book smarts won't necessarily help in those fight-or-flight moments, especially, perhaps, for girls. "I talk to a hundred girls a month who are superassertive, feminist, who can correct their teachers about the symbolism of a novel in class," she said. "Then they're at a party and some dude's hand is on their leg—or between their legs—and they feel like duct tape is over their mouths. They literally can't say, 'Can you move your hand?' Superassertive, but not in that situation, because they're using a different part of themselves. And then there's regret and shame. And that's just because we need to practice." Once again, the room has fallen silent, the kind of hush that occurs when a teacher has truly touched a nerve.

Denison asked for a volunteer, and Jackson, a lanky boy in a Chicago Bulls T-shirt, stood up. "People talk about 'assertive' all the time," Denison said. "And 'aggressive' and 'passive-aggressive.' Those are ways to think about how we react in the real world, especially when we are uncomfortable." She pulled out her cell phone. "So, let's say I borrowed Jackson's phone and said I'd have it back in a day. But it's been three days. Also, I've cracked the screen. Now I'm going to return it to him, and he's going to show us what a passive response would be."

She sauntered over and dumped the device in his hand. "Thanks for the phone, Jackson. It's awesome." She casually gestured to the imaginary crack, adding, "There's just, like, that little thing here."

"No problem," Jackson said.

"Really?" Denison took a step toward him. "Can I borrow it again, then?"

"No, um . . ."

She took another step. "Oh. Well, do you have a car?"

"Yeah, it's over there."

"Can I have the keys?" Jackson pretended to toss them to her, and the scene was over.

Denison turned to the class. "So his fallback was 'I'm uncomfortable, this is unpleasant, I want it to end, and agreeing with her is the fastest way.' But did you see how when I stepped forward and he backed away I was like, 'Yeah, I've got this. I do not have to be accountable in any way. I can take advantage.' So the bummer of that response is 'Am I going to come back?' Oh, *hell* yeah. He's got a Post-it on his forehead with a bull's-eye now. But if he could get to a place of thinking, 'How do I feel right now? What do I think? And what do I want to have happen?' Maybe it would be worth thirty seconds of doing something different so that this obnoxious girl will not come back."

A thin boy in a red-and-white striped shirt raised his hand. "So what exactly is an aggressive fallback?"

"Pushing back against someone before they can go after you," another boy said. "Or saying someone's an asshole."

"Or, like, yelling back if your parents start yelling at you," said one of the girls.

Denison had invited a handful of seniors, students the underclassmen respected, to join her today as de facto teaching assistants. One of them, a girl wearing a fedora and a vintage Violent Femmes T-shirt, raised her hand. "My fallback was, when someone asked, like, 'What do you want to eat?' I'd say, 'Whatever you want.' And I would never, ever, *ever* say what I wanted to do. I've been trying to change that, at least with people I'm comfortable around."

"So it must not have been working for you . . ." Denison prompted.

The girl nodded. "I still do it often. But I didn't like never having the power within myself to say what I wanted. I was always so worried about people not liking me if I said what I might want to do."

"Bringing that back to the situations we're talking about," Denison said, "especially in the hookup scene, especially in the city, where there's so much more access to underage dance clubs—if that's your fallback, it's like a potential minefield of regret. Because there's a lot of ninth-graders who go to those clubs and they say they're not going to drink, they're just going to dance, but they don't think through situations that might make them uncomfortable, or come up with a plan.

"There's a lot of oral sex happening in these clubs, in the back hallways," she continued. "Sometimes it's because people know they don't want to have intercourse, but they haven't practiced saying, 'No, I'm not going to go down on you.' That just seems impossible to them. And then there's a lot of regrettable sexual behavior. And drinking, too. Because you get tired of saying no or getting your mom to text you that you have to come home. So, trying to come up with some real, workable tools, and especially working with your seniors here, people who've been there, is really helpful. And trying to get you guys at the beginning to identify two or three of your fallbacks, even by the end of this semester. Then let's talk about it over the next few years. Let's really work that muscle. To avoid regret and practice that assertiveness is so important. And what's cool is the more you practice the easier it gets."

They acted out a few more passive, aggressive, and assertive scenarios, with Denison urging volunteers to state firmly "How you feel right now, what you think, what you want to have happen." In the few minutes of class time that remained, she fielded several anonymous questions that students had submitted on index cards, then gave out the number for a cell phone she keeps specifically for their calls and texts. Some of her colleagues over the years have questioned Denison's willingness to let kids intrude on her life at any hour of the day or night. "They say it's a boundary issue," she told me later, "but I disagree. I come here and encourage students to question themselves, to name a situation when it's not going well, to acknowledge that it's not going well, and reflect on it. I promise I will advocate for them. If after that I disappeared, if I bailed, I wouldn't be doing my job." Most of the messages she receives are queries about basic facts involving sex and drugs; sometimes they are about relationship dilemmas or the choice between dueling visions of "regret"; sometimes they are just notes of gratitude. Typically they are anonymous, sometimes from friends of friends of students she's taught, kids she'll never meet. Among the texts she had recently received:

"My boyfriend won't touch me after he comes. Is that valid?"

"My girlfriend and I had a mishap with a condom and were wondering if we should get Plan B, but she is on a hormonal birth control pill. Would it be a bad idea to mix medication?"

"I am talking to this guy and he told me (through text), 'You act like you're never gonna suck a dick. That's, like, a girl's job. . . .' We've been 'a thing' for about two months, and I don't know what to do because I want to make a decision on handling this that I won't regret."

"I just took four busses and a train to follow a boy who called me a bitch a month ago and I need to know why I'm here."

"Charis: I so appreciate everything you do. You were such an incredible resource while I was in high school and have motivated me to start a sex ed radio show/podcast at my college!"

Scrolling through the texts, Denison shakes her head in

wonder. "If adults thought about their world and their choices as deeply as the teens who reach out to me do . . ." she said. "They're so thoughtful. Thoughtful before they do something. Thoughtful after they do something. Thoughtful while they do something. It's inspiring."

Going Dutch

Here's a solution for concerned parents: move to the Netherlands. Okay, maybe that's not the most practical advice. Perhaps, though, we can move a little of the Netherlands here. Because the Dutch seem to have it all figured out. While we in the United States have the highest teen pregnancy rate in the industrialized world, they have among the lowest. Our teen birth rate? Eight times higher than theirs, and our teen abortion rate is 1.7 times higher. Yes, there are some significant demographic differences that affect those numbers: we are a more diverse nation than Holland, with higher rates of childhood poverty, fewer social welfare guarantees, and more social conservatives. Yet even when controlling for all that, the difference holds. Consider a study comparing the early sexual experiences of four hundred randomly chosen American and Dutch women at two similar colleges—nearly all white, all middle class, with similar religious backgrounds. So, apples to apples. The American girls had become sexually active at a younger age than the Dutch, had had more encounters with more partners, and were less likely to use birth control. They were more likely to say they'd had first intercourse because of "opportunity" or pressure from friends or partners. In subsequent interviews with some of the participants, the Americans, much like the ones I met, described interactions that were "driven by hormones," in which boys determined relationships, male pleasure was prioritized, and reciprocity was rare. As for the Dutch girls? Their early sexual activity took place in loving, respectful relationships in which they communicated openly with their partners (whom they said they knew "very well") about what felt good and what didn't, about how "far" they wanted to go, and about what kind of protection they would need along the way. They reported more comfort with their bodies and their desires than the Americans and were more in touch with their own pleasure.

It's enough to make you rush out to buy a pair of wooden shoes.

What's their secret? The Dutch girls said that teachers and doctors had talked candidly to them about sex, pleasure, and the importance of a loving relationship. More than that, though, there was a stark difference in how their parents approached those topics. The American moms had focused on the potential risks and dangers of sex, while their dads, if they said anything at all, stuck to lame jokes. Dutch parents, by contrast, had talked to their daughters from an early age about both the joys and responsibilities of intimacy. As a result, one Dutch girl said she told her mother immediately after her first intercourse, "because we talk very open[ly] about this. My friend's mother also asked me how it was, if I had an orgasm and if he had one."

The attitudes of the two nations weren't always so far apart. According to Amy Schalet, an associate professor of sociology at the University of Massachusetts and author of *Not Under My Roof*, in the late 1960s the Dutch, like Americans, roundly disapproved of premarital sex. The sexual revolution transformed attitudes in both countries, but whereas American parents and policy makers responded by treating teen sex as a health crisis, the Dutch went another way: they consciously embraced it as natural, though requiring proper guidance. Their government made pelvic exams, birth control, and abortion free to anyone under twenty-two, with no requirements for parental consent. By the 1990s, when Americans were shoveling millions into the maw of useless abstinence-only education, Dutch teachers (and parents) were busy discussing the positive aspects of sex and relationships, as well as anatomy, reproduction, disease prevention, contraception, and abortion. They emphasized respect for self and others in intimate encounters, and openly addressed masturbation, oral sex, homosexuality, and orgasm. When a Dutch national poll found that most teenagers still believed that boys should be the more active partner during sex, the government added "interaction" skills to its sex ed curricula, such as how to let "the other person know exactly what feels good" and how to set boundaries. By 2005, four out of five Dutch youth said that their first sexual experiences were well timed, within their control, and fun. Eighty-six percent of girls and 93 percent of boys agreed that "We both were equally eager to have it." Compare that to the United States, where two-thirds of sexually experienced teenagers say they wish they had waited longer to have intercourse for the first time.

It's not just about sex, though—according to Schalet, there's a fundamental difference in the two countries' conceptions of how teenagers become adults. American parents consider adolescents to be innately rebellious, in thrall to their "raging hormones." We respond by cracking down on them, setting stringent limits, forbidding or restricting any behavior that might lead to sex or substance use. We end up with a self-fulfilling prophecy: teens assert independence by breaking rules, rupturing their relationships with parents, separating from the family. Sex, which typically involves sneaking around or straight-up lying, becomes a vehicle through which to do that. Charis Denison, for instance, told me that roughly half the questions she fields from students about parents involve how to get contraception or STD testing without Mom and Dad finding out; the other half are on how to bring up sensitive issues so they will actually listen. Both speak to a rift between teenagers and those who love them most—one we parents more or less create. Girls, Schalet said, particularly suffer, wrestling with the incompatibility of remaining a "good daughter" while becoming sexual. They end up either lying to their parents or copping to their behavior but keeping it invisible, outside the home. Either way, closeness can be compromised. Think back to Sam, who said her politically progressive parents behaved "more like a conservative household" where sex was concerned; Megan, who laughingly told me her dad "thinks I'm a virgin"; Holly, whose mother told her "you shouldn't be having sex" when she asked, at age nineteen, to go on the Pill. Each girl was forced to pretend with her parents, to act the innocent. That didn't change her behavior; it just left her unsupported and vulnerable.

Dutch teens, on the other hand, remain closely connected to parents, growing up in an atmosphere of *gezelligheid*, a word most Americans can't even pronounce, but which Schalet translates loosely as "cozy togetherness." Parents and teens are expected to discuss the children's psychological and emotional development, including their burgeoning sexual drives. As part of that, Dutch parents permit—wait for it—sleepovers, which are rare in the United States, except in the most progressive circles. A full two thirds of Dutch teens ages fifteen to seventeen with a steady boy or girlfriend report that the person was welcome to spend the night in their bedrooms. That's not to say it's a free-for-all over there. Quite the opposite: the Dutch actively discourage promiscuity in their children, teaching that sex should emerge from a loving relationship. Negotiating the ground rules for sleepovers, while not always easy (parents admit to a period of "adjustment" and some embarrassment), provides yet another opportunity to exert influence, reinforce ethics, and emphasize the need for protection. Schalet calls it a kind of "soft control." And you can't really argue with the results.

Holland is not perfect. Girls are still more likely than boys to report having been forced to do something sexually. They are more likely to experience pain during sex or have difficulty reaching orgasm. Although they express equal interest to boys in pursuing both lust and love, and can freely admit to sexual desire, Dutch girls who have multiple casual partners or one-night stands do risk being labeled "sluts." Schalet found, though, that the word didn't carry the same sting or stigma that it does in America. The Dutch boys she interviewed, meanwhile, expected to combine sex and love. They said that their fathers had expressly taught them that their partners must be equally up for any sexual activity, that the girls could (and should) enjoy themselves as much as boys, and that, as one boy said, "of course you should not be so stupid to [have sex] with a drunken head." Although she found American boys often yearned for love, too, they tended to consider this a personal quirk, a trait their peers, who were always DTF ("down to fuck"), did not share.

Getting Down and Dirty—and Ethical

"I'm comfortable talking to my parents about sex."

Charis Denison watched as the ninth-graders began to move. Those who agreed with the statement she had just made headed to the north end of the room; those who disagreed went to the south. Denison had made clear that staying in the middle was not an option: the point of this exercise was to force students to take a stand, to defend or maybe even change deeply held beliefs. In this case, however, nearly everyone chose "disagree."

"My parents are weird," one girl explained, seeming to speak for the entire group.

Some of the statements Denison tossed out during this lesson seemed like ringers. When asked, "If a teen does have sex, he or she should use a condom every single time," everyone obviously agreed. Then Denison said, "Oral sex isn't real sex." A few kids tried to stick in the center of the room, but Denison wouldn't let them. "Sometimes in life," she told them, "you have to make a hard choice. You don't get to stay in the middle. Sometimes you just have to bust a move." In the end, the class was divided. "Well," said a girl who had reluctantly disagreed, "it's not *really* sex. But it's not really not-sex, either. It's kind of . . ." She shrugged helplessly. "I don't know."

A boy standing next to her added, "I think you have to be able to get pregnant to be having actual sex."

Denison raised an eyebrow. "So, my thirty-five-year-old friend who is a lesbian and has never been with a guy is a virgin?" she said. At that, the boy looked confused. "No," he said, slowly, "but . . ."

A girl on the "agree" side interrupted. "I think sex is having an intimate moment with someone," she said. "It doesn't have to mean putting something inside of someone." She received several "snaps" of approval for that reply.

Denison's statements became more provocative later, when she repeated this exercise with eleventh-graders. Chaos broke out over whether "a guy going down on a girl is basically the same as a girl going down on a guy." Several students asked Denison,

"It *should* be or it *is*?" but she stayed mum. A handful refused to move from the noncommittal center of the room. Eventually, though, nearly everyone landed in the "agree" camp.

"That's a big group," Denison said, looking them over. "Do you see it playing out that way in reality? Raise your hand if you think that girls are getting as much oral sex as guys." Not a single hand went up. "So I guess we need to talk about what's going on," Denison said.

Next up: "I know someone who has had unwanted sex." Again, nearly everyone landed on the "agree" side of the room.

A boy in a Matchbox Twenty T-shirt raised his hand. "What is 'unwanted'?" he asked. "Is it when you're drunk and you have sex and then the next day you say, 'Ugh, I didn't want that'?"

"Would you call that unwanted sex?" Denison replied.

"Yeah," he said.

A girl in a striped maxi-dress cut in. "But, I think it's kind of unfair to say the guy's a bastard for doing that to you," she said. "If you were like"—she puts on a ditsy, drunken voice—"'Oh that sounds cool!' And then later you go, 'Not cool, dude.' That's not on his plate."

"Does it have to be on someone's plate to feel unwanted?" Denison asked.

The girl shrugged. "No, I guess not."

Denison gestured to the agree side. "People over here: raise your hand if you know more than one person who's had unwanted sex." Most did. "Keep your hand up if you know more than two people who've had unwanted sex." Most hands stayed up. "More than three." Still a lot of hands. "Four." She paused for a long moment. "I'm in love with the teenage population," she finally said. "I think they're the smartest, most creative, most brave population on the planet, but there's a lot of regret going on around this, a lot of confusion and a lot of messiness. What do we need to lessen that? What are we not doing or what do we need to do?"

A boy in a stocking cap raised his hand. "I think that mind-altering substances are called that for a reason. You make decisions under the influence that you wouldn't make sober."

Denison nodded. "Every choice we make we either surrender or gain power, right?" she said. "With alcohol and drugs, you're surrendering power. Which is why people do it sometimes, because they want that. But let's not be ignorant. Let's realize that with each sip, you lose some power to discern what's going on around you; you lose the power to take care of yourself, to judge your emotions."

A girl in a gray sweatshirt chomping on a big wad of bubble gum raised her hand. "I think that you have to make the definition of consent very clear," she said. "If someone doesn't literally say, 'Yes I want to,' then stop. Even if they didn't say no. Even if they're intoxicated. Even if they said they wanted to and then changed their mind. That's not consensual."

"She's saying *make consent clear*," Denison said. "You're making a lot of sense. Someone is hooking up with someone, they're totally into it. The other person is like, 'Is this okay?' And they say, 'Yeah, bring it on!' But then, all of a sudden, it starts *not* being okay. What needs to happen then?"

"The person needs to say, 'I'm not okay with this now,'" the girl said. "'We can either stop or turn it back and do what we were doing.'"

"That's awesome. But what if the person isn't saying it. What could the other person do?"

"Ask if it's okay," the girl replied.

"Excellent," Denison said. "It is super sexy to get consent. The idea of just saying"—she dropped her voice an octave, jutted out her chin like a teenage boy—" 'Hey, is this all right? You okay?' " She paused for a second to let that sink in. "That's *nice*. It's not, 'I would like to take out my legal documentation right now and get my attorney.' " The kids laughed. "And part of it is recognizing that there are a lot of ways to be sexual. It doesn't have to be this linear thing of going from point A to point B. We have all this language, all these metaphors that say you have to go from here to there." She brought up the baseball metaphor, with its familiar images of "rounding the bases," "home runs," and "scoring." "There's never this idea that someone might go up to bat, hit the ball, round second, and say, 'You know what? I kind of like it here. I'm just going to stay here. I'm not going to go all the way home.' You'd lose the game, right? But if someone says yes, that doesn't mean yes all the way through. There's this useful thing around consent: Any good lover is a good listener. And a bad listener is at best a bad lover and at worst a rapist."

The kids gasped. "*Whoa!*" someone said.

"It's about communication," Denison continued. "That doesn't mean you sing 'Kumbaya' in the middle of intercourse, but it does mean you are sharing with your partner. You are being *intimate*. You get to decide what that intimacy looks like and feels like, and you get to define what 'intimate' is. But there are *two* people involved—that 'you' is plural. Another way you can think about it is: 'What will be a positive sexual experience for everyone involved?' "

A boy in a football jersey, both of whose earlobes were pierced, raised his hand. "I never thought of it before, but in that baseball metaphor? You're trying to score *against* them."

"Exactly," Denison agreed. "There's a winner and a loser in baseball. It's a competition."

"So who is supposed to be the loser?" a girl asked. "The other person?"

Denison just smiled.

Watching the kids' interchange reminded me of a conversation I'd had with one of Denison's former students, Olivia, now a freshman in college. Olivia had told me she'd hooked up a lot during ninth and tenth grades. She couldn't say why—she certainly wasn't enjoying herself, and it made her feel, as she put it, "gross." "There wasn't a moment that things changed for me," she said one afternoon as we chatted in a café near her former high school. "I just started to understand that I wasn't behaving how I wanted to behave and I wasn't the person I wanted to be. Charis's class was a huge part of it, though. I learned to consciously make decisions instead of just letting things happen. And I began to really think about my values and my morals." She tugged thoughtfully at a lock of dark hair. "I think the biggest difference is that now I try to live consciously, with intent. Like, I used to think, 'Oh, okay, I guess we're hooking up now,' instead of thinking about whether I really wanted to be doing it. It's not that I stopped hooking up entirely, but by my junior year, I was less impulsive. And I felt very much like I was participating in it, not just going along with it."

TWO TENTH-GRADERS HELD up a poster-size piece of butcher paper with the words "HOOKING UP IS . . ." printed in purple block letters across the top. A few minutes earlier, Denison had handed out markers and had students write responses to phrases she'd penned on a row of similar papers, such as "ABSTINENCE IS . . . ," "SEX IS . . . ," "SEX AND ALCOHOL . . . ," "BEING A VIRGIN IS . . . ," "SLUT SHAMING IS . . . ," "PRUDE

SHAMING IS . . ." They'd broken up into small groups to analyze the results and were now reporting to the class. "We observed that hooking up could be a bunch of different things to a bunch of different people," said a girl whose wavy dark hair fell to her waist. "But it's usually thought of as 'no strings attached' and less complicated. Like something you do at a party." She laughed. "But sometimes it actually turns out to be *more* complicated."

"That's really common for teenagers," Denison said. "You go into a hookup to make things easy, and then sometimes it back-fires. Is that what you're saying? What does that look like?"

"In some cases one person becomes more attached than the other," the girl said, "and believes there's something between them."

"If I were to say the word *hookup*," Denison asked, "how many people, as a gut reaction, see it as a negative thing?" No hands went up. "A positive thing?" Only boys raised their hands. "How many people imagine it just being a thing—not positive or nega-tive but just another choice?" More hands went up, this time split equally between boys and girls.

As they continued the lesson, a number of familiar themes emerged. Although all the responses to "Sex is . . ." were enthusiastic—"In a word," said the tall blond boy who presented for his group, "people think it's 'great!'"—everyone raised a hand when Denison asked who among them knew someone who'd had a negative sexual experience. "Yet there wasn't a sin-gle negative thing on that paper," she mused. "Why do you think that is?" Again, too, they discussed whether oral sex was, indeed, "sex"; only two people agreed that it was, until Denison men-tioned her lesbian friend. "Honestly?" the blond boy said. "Sex should be whatever you want it to be." More snaps.

Over the next hour or so, they discussed their feelings about virginity ("In our group, we didn't like the connotation of 'clean' and 'pure,'" said one of the girls) and abstinence (comments on that had included "sad," "a choice," and "anal"). A boy wearing a basketball jersey sparked a cacophony of responses to the question, "But what *is* abstinence anyway? Is it doing anything but intercourse or is it no contact at all, or what?" The group presenting on sex and alcohol initially suggested, sanctimoniously, that mixing the two was a bad idea. But when Denison asked who knew someone who had hooked up sober, not a single hand went up. Not one. "I'm hearing more and more that nobody gets sexual with someone unless they're in an altered state," she said. "And that can really feed into that regret factor."

"I think in some ways it's easier, though," a girl said. "You can be like, 'Oh, I wasn't thinking. I was drinking.'"

"That's what I call a setup," Denison responded. "Especially for girls: if you're a prude for setting limits and you're a slut if you decide to have sex, then you're screwed no matter what. At least if you get drunk, you can say, 'Well, yeah, I didn't know what I was doing.' So it's a way to not be accountable. And you have to have some empathy around that. It's pretty seductive to be able to have an out of some kind if you're going to be shamed or feel regret either way. So what are you supposed to do? We have to look at that more closely. We'll be talking more about that next time."

In the final moments, as she did every session, Denison answered anonymous questions. Here is a smattering from the classes I observed:

What if I pee during intercourse?

How do you get STDs from oral sex?

Is it true that when girls come, they can squirt fluid halfway across the room?

How big is a normal penis?

How many calories are in sperm?

Does your hymen always break when you lose your virginity?

Do you need lube to give a hand job?

How can I make anal sex feel better to my partner?

Denison answered them matter-of-factly, dispensing facts and correcting myths—including that "everyone" is "doing it." "There's such a perception that everyone is having sex and hooking up," she said, responding to a ninth-grader's concerns, "and that is just not the case. There is such pressure and it's just not that common, especially in ninth grade. There's plenty of people who don't even have their first kiss until at least sophomore year, much less go beyond that. So this notion that someone needs to hook up because it's 'time'?" She shook her head. "We have to really work on that. We have to get back to this idea of 'What am I actually feeling, what do I think about it, what do I want to have happen, and how can I look back without regret?' "

At the same time, she offered this to an eleventh-grader whose friend was having sex with many different people. "Your response doesn't have to be 'That's gross' or 'That's good' or 'That's bad.' You can ask, 'How did that feel to you? What does it bring you? How does it serve you?' Approached in the right way, that can be a great conversation. Then, if you really care about that person, your job is to be their human shield from shame."

There were times, listening to Denison answer those anonymous questions, that I felt a little uncertain. Like when someone in an eleventh-grade class asked how to have intercourse in a way that wouldn't hurt his partner. She talked about easing the penis in and out of the vagina gradually, rather than doing the porn-inspired jack hammer thrust, allowing a girl's body time to acclimate. She suggested a boy could shift his weight so he wasn't always bashing into the same spot, and could "empower" a female partner to grab his hips to control the depth of the penetration. There was no denying it: she was explaining how to have sex. It was the worst nightmare of conservative policy makers realized. Yet this is exactly the kind of discussion that, if Holland is any indication, is needed to combat the pop porn culture, reduce regret, and improve teens' satisfaction when they do choose to have sex (whenever that may be). So what about it makes me cringe? Surely, I'd rather have a daughter in bed with a boy who had a question like this asked and answered than one whose only point of reference was what he'd seen on the Internet. "I am not telling them what to do," Denison would explain to me later. "I am responding to a direct question—one that I get ninety-nine percent of the time, by the way—that rises from a student's respect and sense of accountability to both himself and his partner. If I didn't answer specifically, I'd be a fake, just another adult testing their trust." To the class, she concluded with "It's all about *communication*." And of course she was right.

At the end of each session, Denison pulled several handfuls of condoms from a silver tackle box she carried everywhere with her, sort of like Mary Poppins's carpet bag: it also held the vulva puppet, a model of a penis (nicknamed Richard) for demonstrating proper condom use, individual capsules of personal lubricant, and other tools of her trade. "Keep talking, keep asking questions," she would say. "Knowledge is power." True, I saw a group of boys make a show of scooping up the condoms and tossing them in the air. "Children, be free!" one of them said, laughing. But more often students, both boys and girls, approached respectfully. Some took the condoms casually; others sidled up, pretended to pick up an errant index card or pen, and then subtly slipped one or two condoms into their pockets.

A few kids always hung around as the room emptied, hoping for a private moment with Denison. One girl wanted clarification on the definition of statutory rape. Another wanted to know about Denison's career path so she could emulate it. One afternoon, the last student to approach her was a boy with dark curly hair and wide brown eyes. He ground the toe of his sneaker into the floor as he confided that his girlfriend was pushing to have intercourse, but he wasn't ready. "You'd be surprised at how often boys tell me that," Denison told him. "It must be hard and feel lonely." The boy nodded, his eyes welling up. Denison talked to him for a while, in a voice too low for me to hear. Then she gave him her phone number and e-mail address and told him to feel free to contact her. He nodded and walked away, a little less alone.

THIS BOOK IS about girls, about the ongoing obstacles to their full, healthy sexual expression and the costs of that to their well-being. But I want to leave Denison there, with a boy, because making change has to include them as well. It's no longer enough simply to caution young men against "getting a girl pregnant," or, more likely in the current climate, to warn about the shifting definition of rape. Parents need to discuss the spectrum of pressure, coercion, and consent with their sons, the forces urging them to see girls' limits as a challenge to overcome. Boys need to understand how they, too, are harmed by sexualized media and porn. They need to see models of masculine sexuality that are not grounded in aggression against women, in denigration or conquest. They need to know about shared pleasure, mutuality, reciprocity—to transform from baseball players to pizza eaters. That may not be as hard to do as one might think.

Charis Denison taught mostly high school, so one afternoon I sat in on a week-long coeducational puberty class for fourth- and fifth-graders taught by a pink-haired woman aptly named Jennifer Devine, who was a Unitarian minister as well as a certified sex educator. She spent the first session talking about how, with a few notable differences, puberty was basically the same for what she referred to as "people with vulvas and people with penises": everyone gets taller, everyone gets zits, everyone grows hair in new places, everyone's genitals mature, everyone gets "tingly feelings," everyone becomes capable of making a baby. She also spent a session each on the intricacies of male and female anatomy. After those lessons, she asked students to label drawings of men's and women's reproductive systems, both internal and external, which were rendered with clinical precision. That meant both boys and girls had to name the vulva, the outer and inner labia, the vaginal and urethral openings, the anus. I sat behind two boys, Terrell and Gabe, who were doing fine until suddenly Terrell drew a blank. "Hey, Gabe," he said, pointing to his booklet. "What's this again?" Gabe glanced over. "Oh, that's the clitoris," he replied. "That's for making good feelings."

It's a start.

Parents could learn a thing or two from Gabe. I recently suggested to a friend of mine, a woman who, like me, is a feminist, politically progressive mom of a preteen girl, that it was not enough to teach our daughters about the mechanics of reproduction, not enough to encourage resistance to unwanted sexual pressure, or to tell them that rape is not their fault. It was not even enough to equip them with birth control pills and condoms when the time came. We needed to talk to them about *good* sex, starting with how their own bodies worked, with masturbation and orgasm. She balked. "They don't want to hear about that kind of thing from *us*," she said. No? From where will they hear it, then? They deserve something better than the distorted, false voices that blare at them from TVs, computers, iPhones, tablets, and movie screens. They deserve our guidance rather than our fear and denial in their sexual development. They deserve our help in understanding the dangers that lurk, but also in embracing their desire with respect and responsibility, in understanding the complexities and nuances of sexuality.

After studying the Dutch, Amy Schalet whipped up a four-part "ABCD" model for raising sexually healthy kids. First off, we want them to be autonomous (that's A), to understand desire and pleasure, to be able to assert sexual wishes and set limits, and to prepare responsibly for sexual encounters. Moving slowly, with awareness of desire and comfort, is the best way to gain those skills. Who, after all, is truly more sexually "experienced," a person who has intercourse while drunk to divest herself of virginity or the one who spends three hours kissing a partner, learning about erotic tension, mutual pleasure, intentionality? Frankly, if American parents didn't get any further than A, we'd be ahead of the game.

Nonetheless, there are three more letters. B, for building egalitarian, supportive relationships that value shared interest, respect, care, and trust; C for maintaining and nurturing connection with your child; and D for recognizing the diversity and range of sexual orientation, cultural beliefs, and development among their peers. As for that sleepover? I don't know whether I could get there myself, but I'm not saying never—the argument is awfully compelling. Regardless of how we navigate the details, though, we still can, and must, be more open with our daughters and our sons—and encourage them to be more open with us. My friend is actually wrong: Kids *do* want to hear that from their parents. They really do. In a 2012 survey of over four thousand young people, most said they wish they'd had more information, especially from Mom or Dad, before their first sexual experiences. They particularly wanted to know more from us about relationships and the emotional side of sex. So, think about it: Would you like your teenager to explore and understand her own body thoroughly before plunging ahead with partnered sex? Would you like her notion of what constitutes intimacy to extend beyond intercourse? Would you like her to have fewer partners, and consistently protect herself against disease and pregnancy? How about enjoying her sexual encounters? Transcending gender stereotypes? Would you hope she'll find caring, reciprocal, egalitarian relationships in which she can express her needs and limits? If she does pursue sexual pleasure outside relationships, do you want those experiences, too, to be safe, mutual, and respectful? I know I would. All the more reason to take a deep breath and forge ahead with discussions (that's *multiple* discussions) that include ideas about healthy relationships, communication, satisfaction, joy, mutuality, ethics, and, yes, toe-curling bliss.

After talking to so many girls, I now know what to hope for—for my own daughter and for them. I want sexuality to be a source of self-knowledge and creativity and communication despite its potential risks. I want them to revel in their bodies' sensuality without being reduced to it. I want them to be able to ask for what they want in bed, and to get it. I want them to be safe from disease, unwanted pregnancy, cruelty, dehumanization, and violence. If they are assaulted, I want them to have recourse from their school administrators, employers, the courts. It's a lot to ask for, but it's not too much. We've raised a generation of girls to have a voice, to expect egalitarian treatment in the home, in the classroom, in the workplace. Now it's time to demand that "intimate justice" in their personal lives as well.

Appendix A:
Consent Alone Is a Low Bar for Sex

BY MARY ELIZABETH WILLIAMS

This article first appeared in Salon *at* www.salon.com.
An online version remains in the Salon *archives.*
Reprinted with permission.

I'm not going to tell you to go right now and buy a copy of Peggy Orenstein's *Girls & Sex.* I'm going to tell you to buy two copies: One for yourself, and one for the teenager in your life. Because kids—boys and girls, gay and straight—need to understand not just what a new generation of girls is doing in its intimate life. They need to know what those girls are not doing. Like when they're not saying no to stuff they're not into, because it's easier than arguing about it. Like when they're not asking themselves what feels good—for them. And it's high time, in a cultural moment fraught with sexual panic about hookups and sexting and questions of consent, to shift the conversation—and to fight for young women's right to orgasm.

Peggy Orenstein is a uniquely qualified advocate. As she told me recently in a raucous, *Crazy Ex-Girlfriend*-referencing Skype session, "I feel this really intense connection and resonance with girls. I love talking to girls; I love hanging out with girls. That's why I keep coming back." That she does—Orenstein has spent much of her journalistic career in girl world, from 1994's *School-*

girls: Young Women, Self Esteem, and the Confidence Gap through her 2011 bestseller *Cinderella Ate My Daughter.*

It's been five years since Orenstein's bold, hilarious, and occasionally terrifying foray into the princess industrial complex. Now, the same girls she wrote about then—including those high heels-wearing baby divas—are hitting puberty and beyond, and Orenstein is back to see what happens after growing up with "that 21-piece Disney princess makeup set." Her newest book is an exploration of the lives of high school and college-aged girls today, shown through their various forays into purity balls and walks of shame, into hooking up and coming out. It is not, refreshingly, a condemnation of millennials and their successors—or a hand-wringing call to alarmism. Yes, it discusses frankly the often performative aspects of female adolescent sexuality and doesn't ignore the realities of sexual assault, but *Girls & Sex* refuses to be judgmental or doom and gloom. Instead, it offers something else—a demand for education, enlightenment, and ultimately, the radical notion of equal satisfaction.

When I got your book, I got a copy for myself, and one for my teen daughter. Now she's passed it around among her friends and is having conversations with them. Did you imagine it would be used that way?

I keep hearing that, and that's exactly what I'd hoped. Generally what happens with my books is that the first line is parents, but that it quickly migrates to high school girls and college girls. They can just read it themselves and talk with their friends. It gives them information. And that's power.

I thought, I'm going to give this book to my daughter the summer before she goes into ninth grade. I wrote it because my daughter's going to be going to a gigantic high school where all bets are off, and I was hearing a lot of stuff from my friends whose kids go there. I thought, I need to understand this, and I need to make the world better before my kid gets there.

One of the things that bums me out is seeing young girls who can be so empowered and forthright everywhere else, and then in private they don't even know that they're allowed to want things. They're giving sexual favors and getting nothing in return. You talk about how revelatory it is for these girls when they have an orgasm.

Yeah. One of them said, "I cried. *I cried.*" I mean, that's amazing.

One of my favorite stories is talking to girls about the nonreciprocal thing and saying, "What if guys asked you to get a glass of water, over and over, and they never offered to get you a glass of water? Or if they did, it was totally begrudging?" They would laugh. It's less insulting to be told that you're never going to have reciprocal oral sex and you're always going to be expected to go down on a guy than get him a glass of water.

But we set girls up for that from the get go. Everything in the culture tells them that they are supposed to perform, that they are supposed to pay more attention to being desirable than their own desire. We don't as parents name that whole area between belly button and knees. We don't tell them what a clitoris is. We don't even tell them what a vulva is. We just avoid the whole situation.

[Kids] go into puberty ed class, and female pleasure is not necessary to talk about when you're talking about reproduction. So we don't. We talk about periods. We talk about unwanted pregnancy. And with boys we talk about erections and ejaculation. Then, no surprise, only a third of girls masturbate regularly, only half have ever masturbated. Then we tell them to go into their sexual encounters with a sense of equality. How is that supposed to happen?

We have completely shrouded them. Maybe they figure it out. Maybe they do. But maybe they don't, or maybe they have to get over their early experience, and that's wrong. Look at research: When we talk about sexual satisfaction, we're not talking about the same thing. Young women tend to measure sexual satisfaction by their partner's satisfaction—which is why lesbians are more likely to have orgasms. They're like, "I want you to feel good! No, I want you to feel good!" Heterosexual girls will say, "If he's satisfied, I'm satisfied."

Boys are more likely to measure sexual satisfaction by their own orgasm and their own pleasure. On the flipside, when they talk about bad sex they use completely different language. Boys will say, "I didn't come or I wasn't that attracted to her." Girls will talk about pain and humiliation and degradation. Boys never use that language. We're talking about really different experiences going into it. That's why I love that term "intimate justice." To think about this in terms of equity and power dynamics. Who is entitled to engage sexually? Who is entitled to enjoy? Who is the primary beneficiary?

So what is different now? What is it about this generation that is not just historic "girls pleasing boys" crap that we've all gone through?

Some of it is the same. When I think about that I think, why is it the same? When girls are so much more empowered, when they are so much more vocal—why have things changed so much in the public life and not in the private life? I think all of those things that have grown more intense in an age when the culture has grown so visual and so focused and even more saturated in sexuality. You have a hookup culture where sex precedes intimacy rather than the other way around. And that's not saying, "only have sex in relationships," because that's not true. It's not a moral judgment. We're not saying, "Oh heavens!" We're saying, what are *you* getting out of your sexual experiences? What do *you* want to get from those experiences? What are *you* entitled to, and how do you get there?

What I wanted to do is say, this is what it looks like. This is what you'll probably get out of it. This is what you won't get out of it. Once you know all that, you make your informed choices. Otherwise, the choices are just presented by the media and it's like, we rip off half our clothes, we have intercourse with nothing preceding it. We both have orgasms in three seconds and it's great. Then real girls go into real encounters and think, "What's wrong with me?"

When you have not even had your first kiss, nobody says to you, "It's a symbolic repression that if acted out isn't going to feel particularly good for women." That's part of trying to normalize conversations around sex. It's not about "the talk" when we've never told them they have a vagina and now we're going to tell them about reproduction. It's about talking about these ideas, about rights and entitlement in sexual relationships.

You make it clear that kids who have an abstinence-only education are going to have sex pretty much at around the same time their peers are, and they're going to do with less protection. So what can we learn from religious conservatives about what they're doing right?

I went to a purity ball in Louisiana. I feel like it would be really easy to go to one of those things and just slam them, because the idea behind them is completely wrong. Kids are not going to abstain. We know that maybe they delay sex a little longer. But they have greater rates of pregnancy, they have way higher rates of disease. The boys are six times more likely to engage in anal sex, and both boys and girls are more likely to engage in oral sex and not see that as compromising their virginity. So we know that that's really crap.

That said, there was something really moving about the event, and seeing that the fathers there were having this conversation about their values around sexuality and their expectations around sexuality and their hopes around sexuality with their daughters. It wasn't the conversation I wished they'd have. But when I talked to the girls who were from more typical families, if I asked what their fathers said to them, they just laughed. Their mothers—not always but often—talked to them about risk and danger. Their fathers said nothing. It was almost as if, once we stopped saying, "don't do it until you're married," we didn't know what to replace it with. And we just thought, we aren't going to look.

I didn't like what those fathers were saying. I didn't think it was the right conversation to be having with girls. But I thought at least they're having some conversation and I can't say that that's true among my peers. I also think there's a lot of things our kids don't want to talk about. I think we have to normalize the conversation.

What I try now to do with my daughters [aged sixteen and twelve] is say, "You deserve to be with people who think you're great. Who think you're awesome. That's who deserves your

company." And I talk about people and I don't talk about "boys," because fewer and fewer teenagers identity as exclusively heterosexual. I say that I want the people you date to respect you, to like you, to see how funny you are and I want you to have fun within that. And if it doesn't feel good for you, then there's something wrong.

I always say that this conversation that we're having about consent is so important. But consent is such a low bar for a sexual experience. We've got to do better than that. We have put a lot of emphasis on consent, because we should, but for girls, sometimes they feel, "It ought to feel good because I said yes." And if it's not good, that's confusing and upsetting and hard to understand. We have to say, yes, consent, obviously consent, but consent is the baseline. It's not the experience. We are weird that as a culture we have become more comfortable talking about girls' victimization than girls' pleasure.

I have had a lot of conversations about this over the years with my nieces and my friends' daughters, and a lot of times I would give anything for the earth to swallow me up so I don't have to talk to them about orgasms. But I make myself do it. With one friend's teenage daughter I said, "I'm not going to tell you what you should or shouldn't do. I want you to think about these questions. Do you know where your clitoris is? Have you masturbated? Have you had an orgasm? With yourself? With him? Are you comfortable telling him what you like sexually?" A lot of adult women aren't comfortable with that. If what you're trying to do is express intimacy and mutual pleasure, I'm not sure that rushing to intercourse without understanding that pool of experience is going to get you there. So why are you doing it? I'm not saying it's wrong from a moral perspective, but from the perspective of understanding yourself and your sexuality and exploring, building agency and strength. Ask yourself those questions. It's so important for girls.

What do you think this generation is doing right? What gives you hope for change?

I had this conversation with a girl and she said that all the women in her family were super strong. Then she told me this litany of non-reciprocal experiences and I said, "Why did that happen?" And she told me, "Well, I guess girls are taught to be so meek and deferential." I said, "Wait a second, you just told me how strong you are." She said, "I didn't know that 'strong woman' applied to sex." But then she said, "I'm not doing [other girls] any favors by pretending these things are okay. I'm going to start going into my encounters demanding reciprocity and respect. Otherwise these guys are going to think this is okay, and they're going to keep doing this with somebody else too."

Appendix B:
Girls & Sex by the numbers

COMPILED BY PEGGY ORENSTEIN

The average teenager is exposed to nearly 14,000 references to sex each year on television.[1] Ninety-one percent of comedies and eighty-seven percent of dramas contain sexual content.[2]

Ninety-two percent of the top songs on the Billboard charts are about sex.[3]

Sixty-five percent of girls say selfies boost their confidence; more than half say pictures of them posted by others, however, make them feel bad about their bodies.[4]

Ninety percent of college men and one-third of women viewed porn during the last year.[5]

One in three girls aged fifteen to seventeen say they have performed oral sex on a partner to avoid having intercourse.[6]

More than one-third of teenagers included oral sex in their definition of "abstinence."[7]

The average American first has intercourse at seventeen; by nineteen, three quarters of teens have had sex.[8]

Seventy-two percent of both male and female college students "hook up" at least once by senior year, with the average number of partners being seven.[9] Only one-third of college hookups include intercourse.[10]

Eighty-nine percent of college students get drunk before a random hookup; seventy-five percent get drunk before hooking up with an acquaintance; less than twenty percent drink before a romantic sexual encounter.[11]

The average age of coming out as gay in the United States dropped from twenty-five in 1991 to between fourteen and sixteen today.[12]

Sixteen percent of women 18–24 tried anal sex in 1992. Today 20 percent of women 18–19 have; 40 percent of 20–24 year-olds have.[13]

One in five college women are raped during college;[14] an estimated 11 percent–20 percent report the crime to the authorities.[15]

Fraternity brothers make up 10 percent of accused perpetrators of sexual assault (proportionate to their presence on campus) but they make up 24 percent of repeat offenders.[16] Athletes make up 15 percent of accused assailants (proportionate to their presence on campus) but they made up 20 percent of repeat offenders.[17]

Only thirteen states require that sex education be medically accurate.[18]

Seventy percent of young people 16–24 say they wish they had known more before their first sexual experiences, especially about relationships and the emotional side of sex.[19]

1   "Televised Sexual Content and Parental Mediation," *Media Psychology*, 2009.

2   "Using TV as a Guide." *Journal of Research on Adolescence*, 2006.

3   TheWire.com, 2011.

4   Today.com, 2014.

5   "Generation XXX," *Journal of Adolescent Research*, 2008.

6   Henry Kaiser Family Foundation, 2003.

7   Advocates for Youth, 2002.

8   "Sexual Initiation, Contraceptive Use, and Pregnancy Among Young Adolescents." *Pediatrics*, 2013.

9   "Is Hooking Up Bad For Young Women?" *Contexts* 2010.

10 Ibid.

11 "Do Alcohol and Marijuana Use Decrease the Probability of Condom Use for College Women?" *The Journal of Sex Research*, 2014.

12 *Science Daily*, October 11, 2011.

13 *Sexual Behavior in the United States*, 2012.

14   *Washington Post*/Kaiser Family Foundation Poll, 2015; "What Percent of College Women are Sexually Assaulted in College," *Contexts* 2015; United Educators Insurance, 2015.

15   *Washington Post*/Kaiser Family Foundation Poll, 2015; Department of Justice 2014.

16   United Educators Insurance, 2015.

17   Ibid.

18   "Guttmacher Institute, 2015.

19   *Popular Science*, March 5, 2015.

Acknowledgments

Usually at this point I write about how while authorship is a solitary pursuit, there were many who supported me in it, blah, blah, blah. But that's such a civilized, sanitized way of putting it. What I really mean is this: I am difficult to live with, difficult to be around, difficult to know or interact with in any way while I am engrossed in book writing. The work consumes me. It makes me anxious, obsessive, flaky, and self-absorbed. It makes me grouchy. It makes me emotionally and often physically distant. Sometimes I don't know how those who love me—my friends, my family—can stand it. And yet they do, and that is, for me, the definition of grace.

So let me put it out there for real. For living through this with me yet again, for chewing over the issues, for challenging me, cajoling me, housing me, and enduring me, I would like to thank: Barbara Swaiman, Peggy Kalb, Ruth Halpern, Eva Eilenberg, Ayelet Waldman, Michael Chabon, Sylvia Brownrigg, Natalie Compagni Portis, Ann Packer, Rachel Silvers, Youseef Elias, Stevie Kaplan, Joan Semling Bostian, Mitch Bostian, Judith Belzer, Michael Pollan, Simone Marean, Rachel Simmons, Julia Sweeney Blum, Michael Blum, Danny Sager, Brian McCarthy, Diane Espaldon, Dan Wilson, Teresa Tauchi, Courtney Martin, Moira Kenney, Neal Karlen, ReCheng Tsang Jaffe, Sara Corbett, and Ilena Silverman.

For their assistance with research, I would like to thank Kaela Elias, Sara Birnel-Henderson, Pearl Xu, Evelyn Wang, Henry Bergman, and Sarah Caduto. For acting as sounding boards (and sometimes contending with some very personal questions), thanks to my nieces and nephews, especially Julie Ann Orenstein, Lucy Orenstein, Arielle Orenstein, Harry Orenstein, Matthew Orenstein, and Shirley Kawafuchi. For their guidance, I thank my agent, Suzanne Gluck, my ever-patient editor, Jennifer Barth, as well as Debby Herbenick, Leslie Bell, Patti Wolter, Lucia O'Sullivan, Lisa Wade, Jack Halberstam, Jackie Krasas, Paul Wright, and Bryant Paul. For the luxury of space and time to write uninterrupted, I am profoundly grateful to Peter Barnes and the Mesa Refuge, as well as to the Cindy-licious Ucross Foundation.

Greg Knowles deserves a special place in heaven for rescuing my manuscript when it disappeared into the technological ether. And while looks aren't everything, I sure appreciate what Michael Todd did with mine. Thanks, too, to the staff of *The California Sunday Magazine* and especially Doug McGray for your support and understanding. Special thanks to Charis Denison for all that my reporting put her through.

Most of all, thank you to the generous young women who participated in my interviews and the adults who helped me find them. To protect their privacy, I can't name them here, but you know who you are. It was a pleasure to get to know each and every one of you, and there is no way I could have written this book without you.

Finally, thank you to my family, both extended and immediate. To my husband, Steven Okazaki, so much more love than I could ever express; and to my beloved daughter, Daisy, I hope I haven't embarrassed you too much. I love you boundlessly and wish you the gift of ever and always being fully yourself.

Notes

Introduction: Everything You Never Wanted to Know About Girls and Sex (but Really Need to Ask)

3   The average American has first intercourse: Finer and Philbin, "Sexual Initiation, Contraceptive Use, and Pregnancy Among Young Adolescents."

5   Teen intimacy, it said, ought to be: Haffner, ed., *Facing Facts: Sexual Health for America's Adolescents* .

5   Sara McClelland, a professor of psychology: McClelland, "Intimate Justice."

Chapter 1: Matilda Oh Is Not an Object—Except When She Wants to Be

12   "If they aren't," Moran wrote, "chances are": Moran, *How to Be a Woman*, p. 283.

12   Preschoolers worship Disney princesses: Glenn Boozan, "11 Disney Princesses Whose Eyes Are Literally Bigger Than Their Stomachs," Above Average, June 22, 2015.

12   Self-objectification: American Psychological Association, *Report of the APA Task Force on the Sexualization of Girls*. The groundbreaking report defines sexualization as comprising any one or any combination of the following: "a person's value comes only from his or her sexual appeal or behavior, to the exclusion of other characteristics; a person is held to a standard

that equates physical attractiveness (narrowly defined) with being sexy; a person is sexually objectified—that is, made into a thing for others' sexual use, rather than seen as a person with the capacity for independent action and decision making; and/or sexuality is inappropriately imposed upon a person." See also Madeline Fisher, "Sweeping Analysis of Research Reinforces Media Influence on Women's Body Image," *University of Wisconsin–Madison News*, May 8, 2008.

12 In one study of eighth-graders: Tolman and Impett, "Looking Good, Sounding Good." See also Impett, Schooler, and Tolman, "To Be Seen and Not Heard."

12 Another study linked girls' focus on appearance: Slater and Tiggeman, "A Test of Objectification Theory in Adolescent Girls."

13 A study of twelfth-graders connected self-objectification: Hirschman et al., "Dis/Embodied Voices."

13 Self-objectification has also been correlated: Caroline Heldman, "The Beast of Beauty Culture: An Analysis of the Political Effects of Self-Objectification," paper presented at the annual meeting of the Western Political Science Association, Las Vegas, NV, March 8, 2007. See also Calogero, "Objects Don't Object"; *Miss Representation*, dir. Jennifer Siebel Newsom and Kimberlee Acquaro, San Francisco: Representation Project, 2011.

13 Or, as one alumna put it: Steering Committee on Undergraduate Women's Leadership at Princeton University, *Report of the Steering Committee on Undergraduate Women's Leadership*, 2011; Evan Thomas, "Princeton's Woman Problem," *Daily Beast*, March 21, 2011.

13 "the pressure to look or dress": Liz Dennerlein, "Study: Females Lose Self-Confidence Throughout College," *USA Today*, September 26, 2013.

13 "effortless perfection": Sara Rimer, "Social Expectations Pressuring Women at Duke, Study Finds," *New York Times*, September 24, 2003.

13 It is a commercialized, one-dimensional, infinitely replicated: Levy, *Female Chauvinist Pigs*.

16 rejecting the torture device commonly known as: Haley

Phelan,"Young Women Say No to Thongs," *New York Times*, May 27, 2015.

17   "I will lose weight, get new lenses": Brumberg, *The Body Project*.

18   Comments on girls' pages, too: Steyer, *Talking Back to Facebook*; Fardouly, Diedrichs, Vartanian, et al., "Social Comparisons on Social Media." See also Shari Roan, "Women Who Post Lots of Photos of Themselves on Facebook Value Appearance, Need Attention, Study Finds," *Los Angeles Times*, March 10, 2011; Lizette Borrel, "Facebook Use Linked to Negative Body Image in Teen Girls: How Publicly Sharing Photos Can Lead to Eating Disorders," *Medical Daily*, December 3, 2013; Jess Weiner, "The Impact of Social Media and Body Image: Does Social Networking Actually Trigger Body Obsession in Today's Teenage Girls?" *Dove Self Esteem Project* (blog), June 26, 2013.

19   Their "friends" become an audience: Author's interview with Adriana Manago, Department of Psychology and Children's Digital Media Center, UCLA, May 7, 2010. See also Manago, Graham, Greenfield, et al., "Self-Presentation and Gender on MySpace."

19   Also, especially on photo-sharing sites such as Instagram: Lenhart, "Teens, Social Media and Technology Overview 2015."

19   This despite the fact that 1,499 of the profiles: Bailey, Steeves, Burkell, et al., "Negotiating with Gender Stereotypes on Social Networking Sites."

20   *selfie* was named the "international word of the year": The first recorded use of the word *selfie* was in 2002, in an online chat room by a drunken Australian. It became the word of the year after Oxford's researchers established that its use had spiked 17 percent since the same time in 2012. Ben Brumfield, "Selfie Named Word of the Year in 2013," CNN.com, November 20, 2013.

20   Anyone with a Facebook or Instagram account: Mehrdad Yazdani, "Gender, Age, and Ambiguity of Selfies on Instagram," *Software Studies Initiative* (blog), February 28, 2014.

20   "If you write off the endless stream": Rachel Simmons, "Selfies Are Good for Girls," *Slate DoubleX*, December 1, 2013.

21   But about half also said: Melissa Dahl, "Selfie-Esteem: Teens Say Selfies Give a Confidence Boost," Today.com, February 26, 2014.

21   Body dissatisfaction seems less driven by: Meier and Gray, "Facebook Photo Activity Associated with Body Image Disturbance in Adolescent Girls."

21   the more they look at others' pictures: Fadouly and Vartanian, "Negative Comparisons About One's Appearance Mediate the Relationship Between Facebook Usage and Body Image Concerns." See also Kendyl M. Klein, "Why Don't I Look Like Her? The Impact of Social Media on Female Body Image," CMC Senior Theses, Paper 720, 2013.

21   In 2011 there was a 71 percent increase: Sara Gates, "Teen Chin Implants: More Teenagers Are Seeking Plastic Surgery Before Prom," *Huffington Post*, April 30, 2013.

21   One of every three members: American Academy of Facial Plastic and Reconstructive Surgery, "Selfie Trend Increases Demand for Facial Plastic Surgery," Press release, March 11, 2014. Alexandria, VA: American Academy of Facial Plastic and Reconstructive Surgery.

22   In truth, it's hard to know: Ringrose, Gill, Livingstone, et al., *A Qualitative Study of Children, Young People, and "Sexting."* See also Lounsbury, Mitchell, Finkelhor, et al., "The True Prevalence of 'Sexting.'"

22   one large-scale survey: Englander, "Low Risk Associated with Most Teen Sexting."

22   That's particularly disturbing: Caitlin Dewey, "The Sexting Scandal No One Sees," *Washington Post*, April 28, 2015. This survey of 480 undergraduates found that, for both women and men, being coerced into sexting was more traumatic than being coerced into physical sex.

23   Management consultants use: Roger Schwarz, "Moving from Either/Or to Both/And Thinking," Schwarzassociates.com. If you find you can't make this trick work, try doing it backward, starting with your finger at your waist tracing a counterclockwise circle and moving it up.

23   Deborah Tolman has suggested: Personal conversation, September 20, 2011.

25   Between 2012 and 2013 the number of "Brazilian butt lifts":

American Society of Plastic Surgeons, *2013 Plastic Surgery Statistics Report*, Arlington Heights, IL: American Society of Plastic Surgeons, 2014.

25   "Milk, Milk, Lemonade": "Watch 'Inside Amy Schumer' Tease New Season with Booty Video Parody," *Rolling Stone*, April 12, 2015.

27   As with those pop culture memes: Kat Stoeffel, "bell hooks Was Bored by 'Anaconda,' " *The Cut, New York Magazine* blog, October 9, 2014.

31   "Sexy, but not sexual": Levy, *Female Chauvinist Pigs*.

32   When she was fifteen: Katherine Thomson, "Miley Cyrus on God, Remaking 'Sex and the City,' and Her Purity Ring," *Huffington Post*, July 15, 2008. Other lapsed promise ring Disney Kids include Selena Gomez, Demi Lovato, and the Jonas Brothers. Britney Spears also claimed she was saving intercourse until her wedding night. It turned out she was "not that innocent": she had sex for the first time in high school, years before her famed relationship with Justin Timberlake.

33   Mirroring (and raising further questions about): Foubert, Brosi, Bannon, et al., "Pornography Viewing Among Fraternity Men." See also Bridges, Wosnitzer, Scharrer, et al., "Aggression and Sexual Behavior in Best-Selling Pornography Videos."

34   41 percent of videos also included: Bridges, Wosnitzer, Scharrer, et al., "Aggression and Sexual Behavior in Best-Selling Pornography Videos."

34   Watching natural-looking people: Chris Morris, "Porn Industry Feeling Upbeat About 2014," NBCnews.com, January 14, 2014.

34   nearly 90 percent of 304 random scenes: Bridges, Wosnitzer, Scharrer, et al., "Aggression and Sexual Behavior in Best-Selling Pornography." An earlier study found that whether the director was male or female made little difference in the level of aggression or degradation of women in the films. Chyng, Bridges, Wosnitzer, et al., "A Comparison of Male and Female Directors in Popular Pornography." See also Monk-Turner and Purcell, "Sexual Violence in Pornography," which analyzed a random sample of adult videos and found that most had "sexually violent or dehumanizing/degrading themes." For

example, 17 percent of scenes showed aggression against women, 39 percent of scenes featured female subordination, and 85 percent of scenes showed men ejaculating on women. Barron and Kimmel, "Sexual Violence in Three Pornographic Media," found a progressive increase in sexual violence in pornographic materials from magazines to videos to the Internet.

34   as one eighteen-year-old pursuing: *Hot Girls Wanted*, directed by Jill Bauer and Ronna Gradus, Netflix, 2015.

35   "So when you see consistent depictions of women": Personal interview, Bryant Paul, Indiana University–Bloomington, December 4, 2013.

35   Over 40 percent of children ages ten to seventeen have been exposed: Wolak, Mitchell, and Finkelhor, "Unwanted and Wanted Exposure to Online Pornography in a National Sample of Youth Internet Users." Rates of unwanted or accidental exposure to porn rose from 26 percent in 1999 to 34 percent in 2005, Wolak and her colleagues found.

35   By college, according to a survey of more than eight hundred students: Carroll et al., "Generation XXX."

36   There is some indication that porn has: Regnerus, "Porn Use and Support of Same-Sex Marriage."

36   On the other hand, they're also less likely: Wright and Funk, "Pornography Consumption and Opposition to Affirmative Action for Women." This was true for both men and women, even when controlling for prior attitudes on affirmative action.

36   Among teenage boys, regular porn use: Peter and Valkenburg, "Adolescents' Exposure to Sexually Explicit Online Material and Recreational Attitudes Toward Sex"; Peter and Valkenburg, "The Use of Sexually Explicit Internet Material and Its Antecedents." See also Wright and Tokunaga, "Activating the Centerfold Syndrome"; and Wright, "Show Me the Data!"

36   Porn users are also more likely: Wright and Tokunaga, "Activating the Centerfold Syndrome"; Wright, "Show me the Data!"

36   Male *and* female college students who report recent porn use: Wright and Funk, "Pornography Consumption and Opposition

to Affirmative Action for Women"; Brosi, Foubert, Bannon, et al., "Effects of Women's Pornography Use on Bystander Intervention in a Sexual Assault Situation and Rape Myth Acceptance"; Foubert, Brosi, Bannon, et al., "Pornography Viewing Among Fraternity Men." For a study involving high school students, see Peter and Valkenburg, "Adolescents' Exposure to a Sexualized Media Environment and Notions of Women as Sex Objects."

36   female porn users are less likely than others to: Brosi, Foubert, Bannon, et al., "Effects of Women's Pornography Use on Bystander Intervention in a Sexual Assault Situation and Rape Myth Acceptance." One of the arguments *for* pornography is that rates of sexual assault drop in countries where bans against it have been lifted. But as Paul Wright, a professor of telecommunications at Indiana University–Bloomington, told me, if both male and female porn users are more likely buy into rape myths, if women users are less likely to notice when they are at risk, and if women who are objectified are more likely to be blamed when assaulted, then it may not be that there are fewer rapes in such countries so much as that they go unrecognized or unreported. Author interview, Paul Wright, December 6, 2013.

36   Only 3 percent of females do: Carroll et al., "Generation XXX."

37   They believed that the unnatural thinness: Paul, *Pornified*.

38   "What I'm saying is whether it's rated X": "Joseph Gordon-Levitt, on Life and the Lenses We Look Through," Interview on *Weekend Edition*. National Public Radio, September 29, 2013.

39   And the impact of that garden-variety, "pornified" media: Author interview, Paul Wright, December 6, 2013. See also Fisher, "Sweeping Analysis of Research Reinforces Media Influence on Women's Body Image."

39   The average teenager is exposed to: Fisher et. al., "Televised Sexual Content and Parental Mediation."

39   70 percent of prime-time TV: That's up from 56 percent in 1998, the first year tracked. Ninety-one percent of comedies and 87 percent of dramas contained sexual content, ranging from

innuendo to implied intercourse. Ward and Friedman, "Using TV as a Guide"; Shiver Jr, "Television Awash in Sex, Study Says," *Los Angeles Times*, November 20, 2005.

39 College men who play violent, sexualized video games: Stermer and Burkley, "SeX-Box."

39 College women who, in experiments: Fox, Ralston, Cooper, et al., "Sexualized Avatars Lead to Women's Self-Objectification and Acceptance of Rape Myths"; Calogero, "Objects Don't Object."

39 Meanwhile, in a study of middle and high school girls, those who were: Aligo, "Media Coverage of Female Athletes and Its Effect on the Self-Esteem of Young Women"; Daniels, "Sex Objects, Athletes, and Sexy Athletes."

39 Young women who consume more objectifying media: Calogero, "Objects Don't Object."

39 In other words, as Rachel Calogero: Ibid.

39 The sex in TV and movies: Thirty-five percent of sex on TV occurs between two people who either have never met or are not in a relationship: Kunkel, Eyal, Finnerty, et al., *Sex on TV 4*.

42 Kim's true contribution has been an ingenious "patriarchal bargain": Lisa Wade, "Why Is Kim Kardashian Famous?" *Sociological Images* (blog), December 21, 2010.

43 "Our hopes have gotten so cheesy": Tina Brown, "Why Kim Kardashian Isn't 'Aspirational,'" *Daily Beast*, April 1, 2014.

Chapter 2: Are We Having Fun Yet?

48 "'Do you spit or do you swallow?'": Tamar Lewin, "Teen-Agers Alter Sexual Practices, Thinking Risks Will Be Avoided," *New York Times*, April 5, 1997.

48 The reporter linked that incident: Laura Sessions Stepp, "Unsettling New Fad Alarms Parents: Middle School Oral Sex," *Washington Post*, July 8, 1991.

48 Girls' bodies have always been vectors: Brumberg, *The Body Project*.

49 In 1994, just a few years before: Laumann, Michael, Kolata, et al., *Sex in America*.

49 By 2014, oral sex was so common: Author interview, Debby Herbenick, Indiana University, December 5, 2013.

50   Oral sex practices of minors: Remez, "Oral Sex Among Adolescents."

50   By 2000 the Clinton presidency was winding down: Anne Jarrell, "The Face of Teenage Sex Grows Younger," *New York Times*, April 2, 2000.

50   That was not true: Kann, Kinchen, Shanklin, et al., "Youth Risk Behavior Surveillance—United States, 2013."

50   An article in the now-defunct *Talk* magazine: Linda Franks, "The Sex Lives of Your Children," *Talk*, February 2000; See also Liza Mundy, "Young Teens and Sex: Sex and Sensibility," *Washington Post Magazine*, July 16, 2000.

51   The girl whose color hit farthest down: "Is Your Child Leading a Double Life?" *The Oprah Winfrey Show*. Broadcast October 2003/ April 2004. .

51   A 2004 NBC News/*People* survey taken: Tamar Lewin, "Are These Parties for Real?" *New York Times*, June 30, 2005.

52   By the end of ninth grade: Halpern-Felsher, Cornell, Kropp, and Tschann, "Oral Versus Vaginal Sex Among Adolescents," found that one in five ninth-graders reported having experience with oral sex; 37 percent of boys and 32 percent of girls ages fifteen to seventeen reported oral sex experience; by ages eighteen to nineteen, those numbers had roughly doubled, to 66 percent and 64 percent, respectively. Child Trends DataBank, "Oral Sex Behaviors Among Teens." See also Herbenick et al., "Sexual Behavior in the United States"; Fortenberry, "Puberty and Adolescent Sexuality"; Copen, Chandra, and Martinez, "Prevalence and Timing of Oral Sex with Opposite-Sex Partners Among Females and Males Aged 15–24 Years." Over half of fifteen- to nineteen-year-old girls had oral sex before having first intercourse. See Chandra, Mosher, Copen, et al., "Sexual Behavior, Sexual Attraction, and Sexual Identity in the United States"; Chambers, "Oral Sex"; Henry J. Kaiser Family Foundation, "Teen Sexual Activity," *Fact Sheet*; Hoff, Green, and Davis, "National Survey of Adolescents and Young Adults."

52   Right-wing influence on sex education: Dotson-Blake, Knox, and Zusman, "Exploring Social Sexual Scripts Related to Oral Sex."

52   over a third of teenagers included it: Dillard, "Adolescent Sexual Behavior: Demographics."

52   70 percent agreed that someone: Child Trends DataBank, "Oral Sex Behaviors Among Teens."

53   a widespread belief among teens that it is risk free: Halpern-Felsher, Cornell, Kropp, and Tschann, "Oral Versus Vaginal Sex Among Adolescents." Only 9 percent of teens engaging in oral sex report using a condom. See Child Trends DataBank, "Oral Sex Behaviors Among Teens." See also Copen, Chandra, and Martinez, "Prevalence and Timing of Oral Sex with Opposite-Sex Partners Among Females and Males Aged 15–24 Years."

53   rates of sexually transmitted diseases: Advocates for Youth, "Adolescents and Sexually Transmitted Infections"; See also "A Costly and Dangerous Global Phenomenon." *Fact Sheet.* Advocates for Youth, Washington, DC, 2010; "Comprehensive Sex Education: Research and Result"; Braxton, Carey, Davis, et al., *Sexually Transmitted Disease Surveillance 2013.*

53   The new popularity of oral sex: Steven Reinberg, "U.S. Teens More Vulnerable to Genital Herpes," WebMD, October 17, 2013. See also Jerome Groopman, "Sex and the Superbug," *New Yorker,* October 1, 2012; Katie Baker, "Rethinking the Blow Job: Condoms or Gonorrhea? Take Your Pick," *Jezebel* (blog), September 27, 2012.

53   The number one reason they do it: avoiding STDs ranked fifth for girls in a list of motivations for engaging in oral sex, after improving relationships, popularity, pleasure, and curiosity. It ranked third for boys. Cornell and Halpern-Felsher, "Adolescent Health Brief."

53   For years, psychologists have warned: Gilligan et al., *Making Connections*; Brown and Gilligan, *Meeting at the Crossroads*; Pipher, *Reviving Ophelia.* See also: Simmons, *Odd Girl Out*; Simmons, *The Curse of the Good Girl*; Orenstein, *Schoolgirls.*

54   Boys, incidentally, far and away, said: They were also twice as likely as girls to report feeling good about themselves after oral sex; girls were three times more likely to say they felt used. Brady and Halpern-Felsher, "Adolescents' Reported Consequences of Having

Oral Sex Versus Vaginal Sex." This study specifically looked at
consequences of oral sex among ninth- and tenth-graders.

54 For both sexes, but particularly for girls: Cornell and Halpern-
Felsher, "Adolescent Health Brief."

54 Intercourse could bring stigma, turn you into a "slut": Initiating
fellatio earlier than their peers, however, was associated with
low self-esteem for girls. Fava and Bay-Cheng, "Young Women's
Adolescent Experiences of Oral Sex." Although they said oral
sex was a strategy to gain popularity, ninth- and tenth-grade
girls were only half as likely as boys to feel that strategy was
successful. Brady and Halpern-Felsher, "Adolescents' Reported
Consequences of Having Oral Sex Versus Vaginal Sex"; Cornell
and Halpern-Felsher, "Adolescent Health Brief."

54 the calculus and compromises they made to curry favor: One in
three girls in a national survey of teens reported engaging in
oral sex specifically to avoid intercourse. Hoff, Green, and Davis,
"National Survey of Adolescents and Young Adults."

55 They were both dispassionate and nonpassionate about: Burns,
Futch, and Tolman, " 'It's Like Doing Homework.' "

60 As Anna said, reciprocity in casual encounters: In their research
on college students, Laura A. Backstrom and her colleagues
similarly found that cunnilingus was an assumed part of
relationships but not of hookups. Women who wanted oral sex
in hookups had to be assertive to get it; those who did not want
it were found to be relieved. In relationships, women who did
not want oral sex were uncomfortable, whereas it was considered
a source of pleasure for those who enjoyed it. Backstrom et al.,
"Women's Negotiation of Cunnilingus in College Hookups and
Relationships."

62 Around a third masturbated regularly: According to the
National Survey of Sexual Health and Behavior (NSSHB), more
than three quarters of boys ages fourteen to seventeen say they
have ever masturbated; less than half of girls have; about a
third of girls in every age group masturbate regularly, while the
percentage of boys rises steadily with time. Fortenberry, Schick,
Herbenick, et al., "Sexual Behaviors and Condom Use at Last

Vaginal Intercourse"; Robbins, Schick, Reese, et al., "Prevalence, Frequency, and Associations of Masturbation with Other Sexual Behaviors Among Adolescents Living in the United States of America"; Alan Mozes, "Study Tracks Masturbation Trends Among U.S. Teens," *U.S. News and World Report*, August 1, 2011. According to Caron, *The Sex Lives of College Students*, 65 percent of male students masturbate once a week compared to 19 percent of female students.

63   "Like once you've done *that*, you really must be": Most of the college students Backstrom et al. studied likewise viewed cunnilingus as intimate and emotional, and so more desirable within a relationship. Backstrom et al., "Women's Negotiation of Cunnilingus in College Hookups and Relationships." See also Bay-Cheng, Robinson, and Zucker, "Behavioral and Relational Contexts of Adolescent Desire, Wanting, and Pleasure."

64   They are dirty, the male writer continues: Wayne Nutnot, "I'm a Feminist but I Don't Eat Pussy," Thought Catalog, June 7, 2013.

65   Those early experiences can have a lasting: Schick, Calabrese, Rima, et al., "Genital Appearance Dissatisfaction."

66   Women's feelings about their genitals: Author interview, Debby Herbenick, Indiana University, December 5, 2013; Schick, Calabrese, Rima, et al., "Genital Appearance Dissatisfaction." See also Widerman, "Women's Body Image Self-Consciousness During Physical Intimacy with a Partner."

66   College women in one study who were uncomfortable: Schick, Calabrese, Rima, et al., "Genital Appearance Dissatisfaction." See also Widerman, "Women's Body Image Self-Consciousness During Physical Intimacy with a Partner."

66   Another study, of more than four hundred undergraduates: Bay-Cheng and Fava, "Young Women's Experiences and Perceptions of Cunnilingus During Adolescence."

66   Young women who feel confident: Armstrong, England, and Fogarty, "Accounting for Women's Orgasm and Sexual Enjoyment in College Hookups and Relationships."

69   Largely as a result of the Brazilian trend: American Society for

Aesthetic Plastic Surgery, "Labiaplasty and Buttock Augmentation Show Marked Increase in Popularity," Press release, February 5, 2014; American Society for Aesthetic Plastic Surgery, "Rising Demand for Female Cosmetic Genital Surgery Begets New Beautification Techniques," Press release, April 15, 2013.

69 The most sought-after look: Alanna Nuñez, "Would You Get Labiaplasty to Look Like Barbie?" *Shape*, May 24, 2013. See also Mireya Navarro, "The Most Private of Makeovers," *New York Times*, November 28, 2004.

70 "Thirty percent of female college students say": Herbenick et al., "Sexual Behavior in the United States." The National Survey of Sexual Health and Behavior is the largest survey ever conducted on the sexual practices of men and women ages fourteen to ninety-four.

70 The rates of pain among women: Ibid.

70 In 1992 only 16 percent of women: Herbenick et al., "Sexual Behaviors in the United States." See also: Susan Donaldson James, "Study Reports Anal Sex on Rise Among Teens," ABC. com, December 10, 2008.

70 Girls were expected to endure the act: Bahar Gholipour, "Teen Anal Sex Study: 6 Unexpected Findings," Livescience.com, August 13, 2014.

71 Consider that at every age: Laumann et al., *Sex in America*.

71 Or that girls are four times more: Twelve percent of young women said they tolerate unwanted sexual activity versus 3 percent of young men. Kaestle, "Sexual Insistence and Disliked Sexual Activities in Young Adulthood."

71 According to Sara McClelland, who coined the term: McClelland, "Intimate Justice"; author interview, Sara McClelland, January 27, 2014.

72 For men, it was the opposite: McClelland, "Intimate Justice"; McClelland, "What Do You Mean When You Say That You Are Sexually Satisfied?"; McClelland, "Who Is the 'Self' in Self-Reports of Sexual Satisfaction?"

72 Women's commitment to their partner's satisfaction: In sexual encounters between women, both partners have orgasms 83

percent of the time. Author interview, Lisa Wade, March 19, 2014. See also Douglass and Douglass, *Are We Having Fun Yet?*; Thompson, *Going All the Way.*

Chapter 3: Like a Virgin, Whatever That Is

76   Just last week, Christina told me: In 2012, filmmaker Lina Esco launched a movement called "Free the Nipple," focused on ending the double standard that sexualizes women's upper bodies but not men's. In August 2015, "Go Topless Day" protesters demonstrated in sixty cities across the globe for gender equality in public breast-baring. *Free the Nipple*, dir. Linda Esco New York: IFC Films; Kristie McCrum, "Go Topless Day Protesters Take Over New York and 60 Other Cities for 'Free the Nipple' Campaign,'" *Mirror*, August 24, 2015.

77   Nearly two thirds of teenagers: Sixty-four percent of twelfth-graders have had intercourse at least once. Kann, Kinchen, Shanklin, et al., "Youth Risk Behavior Surveillance—United States, 2013."

77   a sizable number of girls: Seventy percent of sexually experienced girls report their first intercourse was with a steady partner; 16 percent say it was with someone they'd just met or with a friend. Martinez, Copen, and Abma, "Teenagers in the United States: Sexual Activity, Contraceptive Use, and Childbearing, 2006–2010 National Survey of Family Growth."

77   Over half, both in national: Leigh and Morrison, "Alcohol Consumption and Sexual Risk-Taking in Adolescents."

77   Most say they regret their experience: Martino, Collins, Elliott, et al., "It's Better on TV"; Carpenter, *Virginity Lost*. While research doesn't answer the question "Waited for *what*?," Martino and colleagues write that "youth who say they wish they had waited longer to have sex for the first time apparently come to regret their decision to have sex, whether because they felt unprepared for the experience, wish they had shared it with someone else or been at a different point in their relationship, found the sex itself to be unsatisfying, or found that the consequences were not what they hoped or expected they would be."

78   in her book *The Purity Myth*: Valenti, *The Purity Myth.*

81   One in four eighteen-year-olds: Jayson, "More College 'Hookups' but More Virgins, Too."

81   unless they're religious, most don't advertise: Carpenter, *Virginity Lost.*

82   each, more or less, reflected in: Ibid.

83   first intercourse was just a natural: This could be why researcher Sharon Thompson found that young women who recognize and make sexual decisions based on their own desire are more likely than those who ignore or deny it to find pleasure in virginity loss. Thompson, *Going All the Way.*

85   By 2004 more than 2.5 million: Bearman and Brückner, "Promising the Future."

85   I made a note to myself to check: Rector, Jonson, Noyes, et al., *Sexually Active Teenagers Are More Likely to Be Depressed and to Attempt Suicide.*

86   Girls, for instance, are also more likely than boys to be bullied: Dunn, Gjelsvik, Pearlman, et al., "Association Between Sexual Behaviors, Bullying Victimization, and Suicidal Ideation in a National Sample of High School Students."

88   perhaps due to a lack of education, or perhaps: Regnerus, *Forbidden Fruit.* Regnerus found that only half of sexually active teenagers who report seeking guidance from God or the Scriptures when making a tough decision say they use protection every time they have intercourse. Among sexually active youth who say they look to parents or another trusted adult for advice, 69 percent do. Regnerus's findings were drawn from the National Longitudinal Study of Adolescent Health (hereafter Add Health) as well as a national survey he and his colleagues conducted of about 3,400 children ages thirteen to seventeen.

88   Pledging has to feel special: Bearman and Brückner, "Promising the Future." Bearman and Brückner's data were drawn from Add Health.

89   Male pledgers are four times more: Bearman and Brückner, "After the Promise."

89   by their twenties, over 80 percent: Rosenbaum, "Patient Teenagers?"

89   The only lesson that sticks: Ibid.

90   Once wed, they found that talking to friends: Molly McElroy, "Virginity Pledges for Men Can Lead to Sexual Confusion— Even After the Wedding Day," *UW Today*, August 16, 2014.

90   A young woman who had taken: Samantha Pugsley, "It Happened to Me: I Waited Until My Wedding Night to Lose My Virginity and I Wish I Hadn't," *XOJane*, August 1, 2014." See also Jessica Ciencin Henriquez, "My Virginity Mistake: I Took an Abstinence Pledge Hoping It Would Ensure a Strong Marriage. Instead, It Led to a Quick Divorce," *Salon*, May 5, 2013.

90   Meanwhile, a 2011 survey: Darrel Ray and Amanda Brown, *Sex and Secularism*, Bonner Springs, KS: IPC Press, 2011.

92   Again, his concern seemed less: Relationships in middle adolescence have been linked with positive, healthy commitment in later relationships; but they can also sometimes be a symptom of pathology. Like so many of these issues, it depends on the context and the couple. Simpson, Collins, and Salvatore, "The Impact of Early Interpersonal Experience on Adult Romantic Relationship Functioning."

92   What's more, if Dave really: U.S. Census Bureau, "Divorce Rates Highest in the South, Lowest in the Northeast, Census Bureau Reports," News brief, Washington, DC: U.S. Census Bureau, August 25, 2011. See also Vincent Trivett and Vivian Giang, "The Highest and Lowest Divorce Rates in America," *Business Insider*, July 23, 2011.

92   Statistically, the strongest factor: Jennifer Glass, "Red States, Blue States, and Divorce: Understanding the Impact of Conservative Protestantism on Regional Variation in Divorce Rates," Press release, January 16, 2014. Council on Contemporary American Families.

95   even those who believe they've talked: According to a joint poll of readers conducted by *O Magazine* and *Seventeen* that involved a thousand fifteen- to twenty-two-year-olds and a thousand mothers of girls those ages, 22 percent of mothers

believed their daughters were uncomfortable talking to them about sex; 61 percent of girls said they were. The percentage of girls having oral sex (30 percent) was double what mothers knew or suspected. Forty-six percent of girls who'd had intercourse did not tell their mothers. Among the girls who'd had an abortion, many also never told their mothers. Liz Brody, "The *O/Seventeen* Sex Survey: Mothers and Daughters Talk About Sex," *O* Magazine, May 2009. A 2012 Planned Parenthood survey found that while about half of parents said they were comfortable talking about sex with their teenagers, only 19 percent of teens said they were comfortable talking to their parents; and while 42 percent of parents said they'd talked "repeatedly" to their children about sex, only 27 percent of teens agreed. Thirty-four percent said their parents have either talked to them only once or never. Parents in the survey believed they were giving their kids nuanced guidance; the kids were only hearing simple directives, such as "don't." Planned Parenthood. "Parents and Teens Talk About Sexuality: A National Poll," *Let's Talk*, October 2012. See also Planned Parenthood, "New Poll: Parents Are Talking with Their Kids About Sex but Often Not Tackling Harder Issues," Plannedparenthood.org, October 3, 2011.

101 What if, as Jessica Valenti suggests: Valenti, *The Purity Myth*.

Chapter 4: Hookups and Hang-Ups

104 The seismic tectonic shift in premarital sexual behavior really took place: Armstrong, Hamilton, and England, "Is Hooking Up Bad for Young Women?"

104 That's what is meant by the term: Wade and Heldman, "Hooking Up and Opting Out."

105 According to the Online College Social Life Survey: Armstrong, Hamilton, and England, "Is Hooking Up Bad for Young Women?"

105 The behavior is most typical among affluent: Ibid. African American women and Asian men have historically been most marginalized in the sexual marketplace. Gay students, too, have

lower hookup rates, perhaps because, on many campuses, their numbers are small and concerns about safety remain high. See Garcia, Reiber, Massey, et al., "Sexual Hook-Up Culture." According to sociologist Lisa Wade, black students are also more conscious of appearing "respectable" and avoiding stereotypes of the "Mandingo" or "Jezebel." The hookup culture centers around fraternity parties, too, and black frats tend not to have their own houses. Poor and working-class students, often the first in their families to attend college, also avoid the party/hookup scene. Lisa Wade, "The Hookup Elites," *Slate DoubleX*, July 19, 2013.

105 Only a third of these hookups included intercourse: Armstrong, Hamilton, and England, "Is Hooking Up Bad for Young Women?"

105 Kids themselves tend to overestimate: Alissa Skelton, "Study: Students Not 'Hooking Up' As Much As You Might Think," *USA Today*, October 5, 2011; Erin Brodwin, "Students Today 'Hook Up' No More Than Their Parents Did in College," *Scientific American*, August 16, 2013.

105 from the 92 percent of songs on the *Billboard* charts: Dino Grandoni, "92% of Top Ten Billboard Songs Are About Sex," The Wire: News from *The Atlantic*, September 30, 2011.

105 Mindy Kaling, creator and star of: "Not My Job: Mindy Kaling Gets Quizzed on Do-It-Yourself Projects," *Wait, Wait . . . Don't Tell Me!*, National Public Radio, June 20, 2015.

106 The truth is, nearly three quarters: Debby Herbenick, unpublished survey, February 2014.

106 In hookups involving intercourse: Armstrong, England, and Fogarty, "Accounting for Women's Orgasm and Sexual Enjoyment in College Hookups and Relationships."

106 Orgasm may not be the only measure of sexual satisfaction: Ibid.

106 As one boy put it to Armstrong: Ibid.

107 That may partly explain why 82 percent of men said: Garcia, Reiber, Massey, et al., "Sexual Hook-Up Culture." A 2010 study of 832 college students found only 26 percent of women and 50 percent of men reported feeling positive after a hookup. Other studies have found that roughly three quarters of students

regretted at least one previous instance of sexual activity. Owen et al., " 'Hooking up' Among College Students."

107 As the age of first marriage rose: Armstrong, Hamilton, and England, "Is Hooking Up Bad for Young Women?"; Hamilton and Armstrong, "Gendered Sexuality in Young Adulthood."

111 Her schoolwork was suffering, too: The breakup of a relationship is among the most distressing, traumatic events teens report, and evidence is growing that it is a major cause of suicide among youth. Joyner and Udry, "You Don't Bring Me Anything but Down"; Monroe, Rhode, Seeley, et al., "Life Events and Depression in Adolescence."

111 More than half of physical and sexual abuse: According to the CDC, more than one in seven high school girls were physically abused by a romantic partner in the past year, and one in seven was sexually assaulted. Latina and white girls were victims of dating abuse more than black girls. Kann, Kinchen, Shanklin, et al., "Youth Risk Behavior Surveillance—United States, 2013."

111 those experiences prime girls to be victimized again: Exner-Cortens, Eckenrode, and Rothman, "Longitudinal Associations Between Teen Dating Violence Victimization and Adverse Health Outcomes."

114 On most of the campuses I visited: There has recently been some effort to change this as a strategy to reduce sexual assault. Amanda Hess, "Sorority Girls Fight for Their Right to Party," *Slate XXFactor.* January 20, 2015

117 in order to create what Lisa Wade, an associate: Author's interview with Lisa Wade, June 9, 2015.

117 As with intercourse, the proportion of young people: Reductions in rates of binge drinking have been driven by college men, not women. National Center for Chronic Disease Prevention and Health Promotion, "Binge-Drinking: A Serious, Unrecognized Problem Among Women and Girls." See also Rachel Pomerance Berl, "Making Sense of the Stats on Binge Drinking," *U.S. News and World Report,* January 17, 2013.

117 one out of four college women and one out of five: National Center for Chronic Disease Prevention and Health Promotion,

"Binge-Drinking: A Serious, Unrecognized Problem Among Women and Girls." See also Berl, "Making Sense of the Stats on Binge Drinking."

117 Other surveys have found that nearly: "College Drinking," *Fact Sheet*. Kelly-Weeder, "Binge Drinking and Disordered Eating in College Students"; Dave Moore and Bill Manville, "Drunkorexia: Disordered Eating Goes Hand-in-Glass with Drinking Binges," *New York Daily News*, February 1, 2013; Ashley Jennings, "Drunkorexia: Alcohol Mixes with Eating Disorders," ABC News, October 21, 2010.

118 They're most likely to be the most drunk: In one study of men and women who engaged in an uncommitted sexual encounter that included penetrative sex, 71 percent were drunk at the time. Fisher, Worth, Garcia, et al., "Feelings of Regret Following Uncommitted Sexual Encounters in Canadian University Students."

118 a process that involved asceticism: Caitlan Flanagan, "The Dark Power of Fraternities," *Atlantic*, March 2014.

125 the number who fake: Caron, *The Sex Lives of College Students*.

129 Mixing energy drinks with alcohol leaves a person: Centers for Disease Control, "Caffeine and Alcohol," *Fact Sheet*; Linda Carroll, "Mixing Energy Drinks and Alcohol Can 'Prime' You for a Binge," Today.com, *News* (blog), July 17, 2014; Allison Aubrey, "Caffeine and Alcohol Just Make a Wide-Awake Drunk," *Shots: Health News from NPR* (blog), February 11, 2013.

131 But as Armstrong and her colleagues have pointed out: Armstrong, Hamilton, and Sweeney, "Sexual Assault on Campus."

131 Since victims have a hard time: Ibid.

133 A report by the Justice Department released: Thirty-two percent of victims the same age who are *not* in college report their assaults. Laura Sullivan, "Study: Just 20 Percent of Female Campus Sexual Assault Victims Go to Police," *The Two Way*, National Public Radio, December 11, 2014.

137 When they do, boys tend to feel remorse about: Oswalt, Cameron, and Koob, "Sexual Regret in College Students."

Chapter 5: Out: Online and IRL

145 Or maybe writing about male bodies liberates women: For more on fan fiction, see Alexandra Alter, "The Weird World of Fan Fiction," *Wall Street Journal*, June 14, 2012; and Jarrah Hodge, "Fanfiction and Feminism." For a fascinating discussion of why so many "slash" stories are male on male, including those written by lesbians, see Melissa Pittman, "The Joy of Slash: Why Do Women Want It?" *The High Hat*, Spring 2005. In the spring of 2014, Chinese officials arrested twenty authors for the crime of writing male/male slash fiction—most were young women in their twenties. Ala Romano, "Chinese Authorities Are Arresting Writers of Slash Fanfiction," *Daily Dot*, April 18, 2014.

146 The company's policy against posting: It's impossible to know who posts the photos on Reddit, since users are anonymous. Ben Branstetter, "Why Reddit Had to Compromise on Revenge Porn," *Daily Dot*, February 27, 2015.

146 As with their straight peers, the Internet can be: "Girls" includes both cisgender LGB girls, whose self-identified gender conforms to their biological sex, and male-to-female transgender girls. Ditto "boys." See GLSEN, *Out Online*.

146 Yet, LGBTQ kids also turn to the Web for information: LGBT teens are five times as likely as their non-LGBT peers to search for information on sexuality and sexual attraction. They are substantially more likely to have a close online friend. Ibid.

146 More than one in ten disclosed: Ibid. According to a 2012 report by the Human Rights Campaign, *Growing Up LGBT in America*, 73 percent of gay teens are "more honest" about themselves online than in the real world, as opposed to 43 percent of teens who identify as heterosexual—though that, too, seems concerning.

148 the average age of coming out: "Age of 'Coming Out' Is Now Dramatically Younger: Gay, Lesbian and Bisexual Teens Find Wider Family Support, Says Researcher," *Science News*, October 11, 2011.

153 In the early 1990s: Caron, *The Sex Lives of College Students*. According to the Centers for Disease Control, 12 percent of women ages 25–44 report having had a same-sex sexual

encounter in their lifetimes; 6 percent of men do. Chandra, Mosher, Copen, et al., "Sexual Behavior, Sexual Attraction, and Sexual Identity in the United States."

156 In a survey of more than ten thousand: Human Rights Campaign, *Growing Up LGBT in America.*

157 Ryan's organization has linked rejection: Zack Ford, "Family Acceptance Is the Biggest Factor for Positive LGBT Youth Outcomes, Study Finds," ThinkProgress.org, June 24, 2015; Ryan, "Generating a Revolution in Prevention, Wellness, and Care for LGBT Children and Youth."

162 An estimated 0.3 percent of Americans are thought to identify: Gary J. Gates, "How Many People Are Lesbian, Gay, Bisexual, and Transgender?" April 2011, Williams Institute on Sexual Orientation Law and Public Policy, UCLA School of Law, Los Angeles.

162 About 3.5 percent of adults identify as gay: Gates, "How Many People Are Lesbian, Gay, Bisexual, and Transgender?" See also Gary J. Gates and Frank Newport, "Special Report: 3.4% of U.S. Adults Identify as LGBT," poll, Gallup.com, October 8, 2012. 4.6 percent of men and 8.3 percent of women ages eighteen to twenty-nine identify as LGBT, the highest rate of any age group. The American public, when polled, believed that 23 percent of adults were gay: Frank Newport, "Americans Greatly Overestimate Percent Gay, Lesbian in U.S," poll, Gallup.com, May 21, 2015.

162 The true number is hard to quantify: There is some dispute over whether "genderqueer" individuals are transgender, or vice versa. Gates, "How Many People Are Lesbian, Gay, Bisexual, and Transgender?"

162 They fell in love in a support group for transgender teens: They have since broken up, and each plans to publish a memoir. Janine Radford Rubenstein, "Arin Andrews and Katie Hill, Transgender Former Couple, to Release Memoirs," *People*, March 11, 2014.

163 They may replace *he* and *she* with: For a rundown of gender-neutral pronouns and their meanings, see "The Need for a Gender Neutral Pronoun," *Gender Neutral Pronoun Blog*, January 24, 2010.

See also Margot Adler, "Young People Push Back Against Gender Categories."

163 her parents said their first inkling: Solomon P. Banda and Nicholas Riccardi, "Coy Mathis Case: Colorado Civil Rights Division Rules in Favor of Transgender 6-Year-Old in Bathroom Dispute," Associated Press, June 24, 2013; Sabrina Rubin Erdely, "About a Girl: Coy Mathis' Fight to Change Gender," *Rolling Stone*, October 28, 2013.

Chapter 6: Blurred Lines, Take Two

168 They were disdainful of girls and female teachers: For an outstanding account of the Glen Ridge rape and its impact, see Lefkowitz, *Our Guys*.

169 It wouldn't be until 2015 that Tyson's former manager: Nicholas Godden, " 'Mike Tyson Rape Case Was Inevitable, I'm Surprised More Girls Didn't Make Claims Against Him,'" *Mail Online*, February 9, 2015.

171 "They would say, 'Yes, I held' ": Kamenetz, "The History of Campus Sexual Assault."

172 Other media outlets: *The Date Rape Backlash Media and the Denial of Rape*, Jhally, prod.

172 When Roiphe lost her novelty: Zoe Heller, "Shooting from the Hip," *Independent*, January 17, 1993.

172 whose book *Who Stole Feminism*: Hoff Sommers, *Who Stole Feminism*.

176 Using the narrowest definition of rape: Raphael, *Rape Is Rape*.

176 Still, given that according to the Census Bureau: There was a total of more than 5.7 million female undergraduates at four-year institutions and more than 3.8 million at two-year institutions. U.S. Census Bureau, *School Enrollment in the United States 2013*, Washington, DC: U.S. Census Bureau, September 24, 2014.

176 The Association of American Universities': Cantor, Fisher, Chibnall, et al., *Report on the AAU Campus Climate Survey on Sexual Assault and Sexual Misconduct*.

177 and 25 percent reported at least one: Ford and England, "What Percent of College Women Are Sexually Assaulted in College?"

A third survey, released in 2015 by United Educators, which
provides liability insurance to schools, found that 30 percent
of rapes reported at its 104 client schools between 2011 and
2013 were committed through force or threat of force and 33
percent were committed while the victim was incapacitated. In
another 13 percent of cases, the perpetrator didn't use force, but
continued engaging in sexual contact after the victim hesitated
or verbally refused. Eighteen percent of cases were labeled
"failed consent": the perpetrator used no force, threat of force,
or coercion but "ignored or misinterpreted cues or inferred
consent from silence or lack of resistance." The remaining 7
percent of rapes involved the use of a knockout drug. Ninety-
nine percent of perpetrators were male. Claire Gordon, "Study:
College Athletes Are More Likely to Gang Rape," *Al Jazeera
America*, February 26, 2015.

177 that brings us back to one in four: Another 2015 study, of 483
students at an unnamed private university in upstate New York,
found that 18.6 percent of freshman women were victims of
rape or attempted rape. Carey, Durney, Shepardson, et al.,
"Incapacitated and Forcible Rape of College Women."

177 it's probably not surprising that by 2006: Kristen Lombardi,
"Campus Sexual Assault Statistics Don't Add Up," Center for
Public Integrity, December 2009. Between 2009 and 2014, over
40 percent of schools in a national sample had not conducted
a single assault investigation. United States Senate, U.S. Senate
Subcommittee on Financial and Contracting Oversight, *Sexual
Violence on Campus*.

178 lower burden of proof: Michael Dorf, " 'Yes Means Yes' and
Preponderance of the Evidence," *Dorf on Law* (blog), October 29,
2014.

179 Among them were the most prestigious in the country: Edwin
Rios, "The Feds Are Investigating 106 Colleges for Mishandling
Sexual Assault. Is Yours One of Them?" *Mother Jones*, April 8,
2015.

180 appears to reflect a new willingness: "New Education
Department Data Shows Increase in Title IX Sexual Violence

Complaints on College Campuses," Press release, May 5, 2015,
Office of Barbara Boxer, U.S. Senator, California.

180 Afterward, though, they sink back: Yung, "Concealing Campus
Sexual Assault."

180 Twenty-eight percent of female college freshmen in a 2015:
Unlike some other surveys, this one limited itself to the legal
definition of rape; it did not include forced fondling or forced
kissing. Carey et al., "Incapacitated and Forcible Rape of
Women." The U.S. Justice Department has found that nearly one
in five girls ages fourteen to seventeen had been the victims of
attempted or completed assault. Finkelhor, Turner, and Ormrod,
"Children's Exposure to Violence."

183 "You don't think you ruined my life forever?": Jason Riley and
Andrew Wolfson, "Louisville Boys Sexually Assaulted Savannah
Dietrich 'Cause We Thought It Would Be Funny,'" *Courier
Journal*, August 30, 2012.

185 often both victim and assailant: Krebs, Lindquist, and Warner,
The Campus Sexual Assault (CSA) Study Final Report.

185 Yet in 2013, when Emily Yoffe wrote on *Slate DoubleX*: Emily Yoffe,
"College Women: Stop Getting Drunk," *Slate DoubleX*, October
15, 2013.

186 Women metabolize liquor differently from men, too: Centers for
Disease Control, "Binge Drinking: A Serious Under-Recognized
Problem Among Women and Girls."

186 If you really want to reduce assault, they said, wouldn't it be
equally: Gordon, "Study: College Athletes Are More Likely
to Gang Rape"; Abbey, "Alcohol's Role in Sexual Violence
Perpetration"; Davis, "The Influence of Alcohol Expectancies
and Intoxication on Men's Aggressive Unprotected Sexual
Intentions"; Foubert, Newberry, and Tatum, "Behavior
Differences Seven Months Later"; Carr and VanDeusen, "Risk
Factors for Male Sexual Aggression on College Campuses";
Abbey, Clinton-Sherrod, McAuslan, et al., "The Relationship
Between the Quantity of Alcohol Consumed and Severity of
Sexual Assaults Committed by College Men"; Norris, Davis,
George, et al., "Alcohol's Direct and Indirect Effects on

Men's Self-Reported Sexual Aggression Likelihood"; Abbey et al., "Alcohol and Sexual Assault"; Norris et al., "Alcohol and Hypermasculinity as Determinants of Men's Empathic Responses to Violent Pornography."

186 It lowers their inhibition; it allows them to disregard: Abbey, "Alcohol's Role in Sexual Violence Perpetration"; Davis, "The Influence of Alcohol Expectancies and Intoxication on Men's Aggressive Unprotected Sexual Intentions"; Abbey et al., "Alcohol and Sexual Assault."

186 By contrast, sober guys not only are less sexually coercive: Abbey, "Alcohol's Role in Sexual Violence Perpetration"; Orchowski, Berkowitz, Boggis, et al., "Bystander Intervention Among College Men."

187 Six hundred thousand students ages eighteen to twenty-four: Nicole Kosanake and Jeffrey Foote, "Binge Thinking: How to Stop College Kids from Majoring in Intoxication," *Observer*, January 21, 2015.

187 In that same two-month period: Dan Noyes, "Binge Drinking at UC Berkeley Strains EMS System," *Eyewitness News*, ABC, November 7, 2013; Emilie Raguso, "Student Drinking at Cal Taxes Berkeley Paramedics," Berkeleyside.com, November 12, 2013; Nico Correia, "UCPD Responds to 8 Cases of Alcohol-Related Illness Monday Morning," *Daily Californian*, August 26, 2013. In 2012, twelve students were transported to the hospital during the first two weeks of school at UC Berkeley; in 2011 there were eleven incidents in the month of August alone. In 2014, however, the number of incidents during the first weekend of school dropped by half. *Daily Californian*, "Drinking Is a Responsibility," August 26, 2014.

187 And yet when binge drinking rises, so does sexual assault: Mohler-Kuo, Dowdall, Koss, et al., "Correlates of Rape While Intoxicated in a National Sample of College Women." This is not, again, to say alcohol causes rape, but that rapists use alcohol in a variety of ways to abet their crimes.

188 "Who knows what their intentions were?": Noyes, "Binge Drinking at UC Berkeley Strains EMS System."

188 nearly 60 percent are unsure: "Poll: One in 5 Women Say They Have Been Sexually Assaulted in College," *Washington Post*, June 12, 2015.

193 " 'we need to stick together and prevent shit like this' ": André Rouillard, "The Girl Who Ratted," *Vanderbilt Hustler*, April 16, 2014.

194 Although, oddly, as criminologist Jan Jordan has pointed out: Raphael, *Rape Is Rape*.

194 "Rape recantations could be the result of the complainants' ": Additionally, victims were urged to take a polygraph test, a practice that has since been abandoned as adversely affecting their willingness to come forward. Rape victims asked to take a polygraph test believe they are being doubted from the get-go. Kanin, "False Rape Allegations."

194 They place false claim rates at between 2 and 8 percent: Raphael, *Rape Is Rape*; Lisak, Gardinier, Nicksa, et al., "False Allegations of Sexual Assault: An Analysis of Ten Years of Reported Cases."

194 Certainly it is important to bear in mind the potential for false claims: Sinozich and Langton, *Special Report: Rape and Sexual Assault Victimization Among College-Age Females, 1995–2013*; Tyler Kingkade, "Fewer Than One-Third of Campus Sexual Assault Cases Result in Expulsion," *Huffington Post*, September 29, 2014; Nick Anderson, "Colleges Often Reluctant to Expel for Sexual Violence," *Washington Post*, December 15, 2014.

195 Emily Yoffe, who also raises the specter: Emily Yoffe, "The College Rape Overcorrection," *Slate DoubleX*, December 7, 2014.

195 "we are also teaching a generation of young women": Emily Yoffe, "How *The Hunting Ground* Blurs the Truth," *Slate DoubleX*, February 27, 2015.

195 Young women, as I've said, remain: See Tolman, Davis, and Bowman, "That's Just How It Is."

197 Another risk-reduction program: Senn, Eliasziw, Barata, et al., "Efficacy of a Sexual Assault Resistance Program for University Women." This is particularly important because rapists target freshman women. The resistance program involved four three-hour units in which skills were taught and practiced.

The goal was for young women to be able to assess risk from acquaintances, overcome emotional barriers in acknowledging danger, and engage in effective verbal and physical self-defense.

198 "I wanted to not cause a conflict": Bidgood, "In Girl's Account, Rite at St. Paul's Boarding School Turned into Rape."

198 though that percentage dropped to 13.6 percent: Edwards et al., "Denying Rape but Endorsing Forceful Intercourse."

198 it also maintains sexual availability as: Katha Pollitt, "Why Is 'Yes Means Yes' So Misunderstood?" *Nation*, October 8, 2014.

203 "Good girlfriends" say yes: Laina Y. Bay-Cheng and Rebecca Eliseo-Arras, "The Making of Unwanted Sex: Gendered and Neoliberal Norms in College Women's Unwanted Sexual Experiences," *Journal of Sex Research* 45, no. 4 (2008): 386–97.

203 What, these young people wondered: For some college women, complying with unwanted sex may be a reaction to having refused in the past and then been coerced by their partner. In a study of undergraduates, a woman was seven times more likely to have engaged in sexual compliance if her partner had previously coerced or assaulted her. Katz and Tirone, "Going Along with It."

204 fraternity brothers and athletes are disproportionately: In 2015, United Educators, which offers liability insurance to schools, released an analysis of 305 sexual assault reports from 104 client colleges made between 2011 and 2013. Although 10 percent of accused perpetrators were fraternity brothers (proportionate to their presence on campus), they made up 24 percent of repeat offenders; 15 percent of accused assailants were athletes, also proportionate to their presence on campus, yet they made up 20 percent of repeat offenders. Athletes were also three times more likely than other students to be involved in gang assaults, committing 40 percent of multiple perpetrator attacks reported to schools. Gordon, "Study: College Athletes Are More Likely to Gang Rape."

Chapter 7: What If We Told Them the Truth?
207 and a good-faith effort to satisfy everyone involved: Abraham, "Teaching Good Sex."

208 Unmarried women could not legally procure contraception: Luker, *When Sex Goes to School.*

208 Although, even then, over half of women: Schalet, *Not Under My Roof.* As late as 1969, two thirds of Americans thought it was wrong to have sexual relations before marriage. Lydia Saad, "Majority Considers Sex Before Marriage Morally Okay," Gallup News Service, May 24, 2001.

208 As sex became untethered: By the early 1970s the disapproval rate for premarital sex had dropped to 47 percent. By 1985, more than half of Americans agreed that premarital sex was "morally okay." Saad, "Majority Considers Sex Before Marriage Morally Okay." In 2014, 66 percent of Americans felt sex between an unmarried man and woman was "largely acceptable." Rebecca Riffkin, "New Record Highs in Moral Acceptability," poll, Gallup.com, May 2014.

209 They argued that the "epidemic" of teen motherhood: Moran, *Teaching Sex.*

209 "To reassure them that infidelity is widespread?": Moran, *Teaching Sex.*

210 "the social, psychological and health gains": Ibid.

210 by 1999, 40 percent of those supposedly teaching comprehensive sex ed: Ibid.

210 By 2003, 30 percent of public school sex: U.S. House of Representatives, *The Content of Federally Funded Abstinence-Only Education Programs.*

211 that money might just as well have been set on fire: Nicole Cushman and Debra Hauser, "We've Been Here Before: Congress Quietly Increases Funding for Abstinence-Only Programs," *RH Reality Check*, April 23, 2015.

211 Studies stretching back over a decade have found: U.S. House of Representatives, *The Content of Federally Funded Abstinence-Only Education Programs*; Hauser, "Five Years of Abstinence-Only-Until-Marriage Education: Assessing the Impact," 2004, Advocates for Youth, Washington, DC; Kirby, "Sex and HIV Programs"; Trenholm, Devaney, Fortson, et al., "Impacts of Four Title V, Section 510 Abstinence Education Programs."

211 They are, however, a lot more likely to become: Kohler,

Manhart, and Lafferty, "Abstinence-Only and Comprehensive Sex Education and the Initiation of Sexual Activity and Teen Pregnancy."

211 otherwise they would have given it up long ago for: Amanda Peterson Beadle, "Teen Pregnancies Highest in States with Abstinence-Only Policies," ThinkProgress, April 10, 2012; Rebecca Wind, "Sex Education Linked to Delay in First Sex," Media Center, Guttmacher Institute, March 8, 2012; Advocates for Youth, "Comprehensive Sex Education"; and "What Research Says About Comprehensive Sex Education."

211 $185 million earmarked for research and programs that: This discretionary funding stream includes $110 million for the president's Teen Pregnancy Prevention Initiative (TPPI), which is under the jurisdiction of the Office of Adolescent Health, and $75 million for the Personal Responsibility Education Program (PREP), which was part of the Affordable Care Act. See "A Brief History of Federal Funding for Sex Education and Related Programs."

212 Meanwhile, $75 million in abstinence-only funds: "Senate Passes Compromise Bill Increasing Federal Funding for Abstinence-Only Sex Education," *Feminist Majority Foundation: Feminist Newswire* (blog), April 17, 2015.

212 What this means for parents is that you never know: For information on individual state requirements as of 2015, see Guttmacher Institute, "Sex and HIV Education."

212 a judge ruled for the first time against: Bob Egeiko, "Abstinence-Only Curriculum Not Sex Education, Judge Rules," *San Francisco Chronicle*, May 14, 2015. A 2011 study conducted by UC San Francisco found uneven compliance with California state laws on sex education. In a sampling of California school districts, more than 40 percent failed to teach about condoms and other contraceptive methods in middle school; and in high school, 16 percent of students were taught that condoms were ineffective, and 70 percent of districts failed to comply with provisions of the law that require age-appropriate materials about sexual orientation.

Sarah Combellick and Claire Brindis, *Uneven Progress: Sex Education in California Public Schools*, November 2011, San Francisco: University of California–San Francisco Bixby Center for Global Reproductive Health.

213 "TAKE CARE & HAVE FUN": Alice Dreger, "I Sat In on My Son's Sex-Ed Class, and I Was Shocked by What I Heard," *The Stranger*, April 15, 2015; Sarah Kaplan, "What Happened When a Medical Professor Live-Tweeted Her Son's Sex-Ed Class on Abstinence," *Washington Post*, April 17, 2015.

219 Consider a study comparing the early sexual experiences: Brugman, Caron, and Rademakers, "Emerging Adolescent Sexuality."

220 "My friend's mother also asked me how it was": Ibid.

220 they consciously embraced it as natural: Schalet, *Not Under My Roof*. See also Saad, "Majority Considers Sex Before Marriage Morally Okay." Gallup did not measure Americans' attitudes toward teen sex in particular until 2013, when it found a significant age discrepancy in beliefs. Only 22 percent of adults over fifty-five agreed that sex between teenagers was "morally acceptable," whereas 30 percent of adults thirty-five to fifty-four and 48 percent of those eighteen to thirty-four agreed it was. Joy Wilke and Lydia Saad, "Older Americans' Moral Attitudes Changing," poll, Gallup.com, May 2013.

221 When a Dutch national poll found that most teenagers still: Schalet, *Not Under My Roof*.

221 Compare that to the United States, where two thirds: Martino, Collins, Elliott, et al., "It's Better on TV."

222 Either way, closeness can be compromised: Schalet, *Not Under My Roof*.

223 Girls are still more likely than boys to report: Vanwesenbeeck, "Sexual Health Behaviour Among Young People in the Netherlands."

223 Dutch girls who have multiple casual partners: Schalet, *Not Under My Roof*.

235 After studying the Dutch: Schalet, "The New ABCD's of Talking About Sex with Teenagers."

236 They particularly wanted to know more from us about:
Alexandra Ossola, "Kids Really Do Want to Have 'The Talk' with
Parents," *Popular Science*, March 5, 2015.

236 All the more reason to take a deep breath: Schear, *Factors That
Contribute to, and Constrain, Conversations Between Adolescent Females
and Their M—others About Sexual Matters*. William Fisher, a professor
of psychology as well as obstetrics and gynecology, found that teens
who felt positively about sex were more likely to use contraception
and disease protection. They were also more likely to communicate
with their partner. Fisher, "All Together Now."

Selected Bibliography

Abbey, Antonia. "Alcohol's Role in Sexual Violence Perpetration: Theoretical Explanations, Existing Evidence, and Future Directions." *Drug and Alcohol Review* 30, no. 5 (2011): 481–89.

Abbey, Antonia, A. Monique Clinton-Sherrod, Pam McAuslan, et al. "The Relationship Between the Quantity of Alcohol Consumed and Severity of Sexual Assaults Committed by College Men." *Journal of Interpersonal Violence* 18, no. 7 (2003): 813–33.

Abbey, Antonia, Tina Zawacki, Philip O. Buck, and A. Monique Clinton. "Alcohol and Sexual Assault." *Alcohol Research and Health* 25, no. 1 (2001): 43–51.

Abraham, Laurie. "Teaching Good Sex." *New York Times Magazine*, November 16, 2011.

Advocates for Youth. "Comprehensive Sex Education: Research and Results." *Fact Sheet*. Washington, DC, 2009.

Aligo, Scott. "Media Coverage of Female Athletes and Its Effect on the Self-Esteem of Young Women." *Research Brief: Youth Development Initiative* 29, September 15, 2014, Texas A&M University, College Station, TX.

Allison, Rachel, and Barbara J. Risman. "A Double Standard for 'Hooking Up': How Far Have We Come Toward Gender Equality?" *Social Science Research* 42, no. 5 (2013): 1191–206.

———. " 'It Goes Hand in Hand with the Parties': Race, Class, and Residence in College Student Negotiations of Hooking Up." *Sociological Perspectives* 57, no. 1 (2014): 102–23.

American Psychological Association. *Report of the APA Task Force on the Sexualization of Girls.* Washington, DC: American Psychological Association, 2007.

American Sociological Association. "Virginity Pledges for Men Can Lead to Sexual Confusion—Even After the Wedding Day." *Science Daily,* August 17, 2014.

Anonymous. "The Pretty Game: Objectification, Humiliation and the Liberal Arts." *Bowdoin Orient,* February 13, 2014.

Armstrong, Elizabeth A., Paula England, and Alison C. K. Fogarty. "Accounting for Women's Orgasm and Sexual Enjoyment in College Hookups and Relationships." *American Sociological Review* 77 (2012): 435–62.

Armstrong, Elizabeth A., Laura Hamilton, and Paula England. "Is Hooking Up Bad for Young Women?" *Contexts* 9, no. 3 (2010): 22–27.

Armstrong, Elizabeth, and Brian Sweeney. "Sexual Assault on Campus: A Multilevel, Integrative Approach to Party Rape." *Social Problems* 53, no. 4 (2006): 483–99.

Backstrom, Laura, et al. "Women's Negotiation of Cunnilingus in College Hookups and Relationships." *Journal of Sex Research* 49, no. 1 (2012): 1–12.

Bailey, Jane, Valerie Steeves, Jacquelyn Burkell, et al. "Negotiating with Gender Stereotypes on Social Networking Sites: From 'Bicycle Face' to Facebook." *Journal of Communication Inquiry* 37, no. 2 (2013): 91–112.

Barron, Martin, and Michael Kimmel. "Sexual Violence in Three Pornographic Media: Toward a Sociological Explanation." *Journal of Sex Research* 37, no. 2 (2000): 161–68.

Bay-Cheng, Laina Y., and Nicole M. Fava. "Young Women's Experiences and Perceptions of Cunnilingus During Adolescence." *Journal of Sex Research* 48, no. 6 (2010): 531–42.

Bay-Cheng, Laina Y., Adjoa D. Robinson, and Alyssa N. Zucker. "Behavioral and Relational Contexts of Adolescent Desire, Wanting, and Pleasure: Undergraduate Women's Retrospective Accounts." *Journal of Sex Research* 46 (2009): 511–24.

Bearman, Peter S., and Hanna Brückner. "Promising the Future: Virginity Pledges and First Intercourse." *American Journal of Sociology* 106, no. 4 (2001): 859–912.

Bersamin, Melina, Deborah A. Fisher, Samantha Walker, Douglas L. Hill, et al. "Defining Virginity and Abstinence: Adolescents' Interpretations of Sexual Behaviors." *Journal of Adolescent Health* 41, no. 2 (2007): 182–88.

Bersamin, Melina, Samantha Walker, Elizabeth. D. Walters, et al. "Promising to Wait: Virginity Pledges and Adolescent Sexual Behavior." *Journal of Adolescent Health* 36, no. 5 (2005): 428–36.

Bisson, Melissa A., and Timothy R. Levine. "Negotiating a Friends with Benefits Relationship." *Archives of Sexual Behavior* 38 (2009): 66–73.

Black, M. C., K. C. Basile, M. J. Breiding, et al. *The National Intimate Partner and Sexual Violence Survey (NISVS): 2010 Summary Report.* Atlanta: National Center for Injury Prevention and Control, Centers for Disease Control and Prevention, 2011.

Bonino, S. et al. "Use of Pornography and Self-Reported Engagement in Sexual Violence Among Adolescents." *European Journal of Developmental Psychology* 3 (2006): 265–88.

Brady, Sonya S., and Bonnie L. Halpern-Felsher. "Adolescents' Reported Consequences of Having Oral Sex Versus Vaginal Sex." *Pediatrics* 119, no. 2 (2007): 229–36.

Bridges, Ana J., Robert Wosnitzer, Erica Scharrer, et al. "Aggression and Sexual Behavior in Best-Selling Pornography Videos: A Content Analysis Update." *Violence Against Women* 16, no. 10 (2010): 1065–85.

Brixton, James, Delicia Carey, Darlene Davis, et al. *Sexually Transmitted Disease Surveillance, 2013.* Atlanta: Centers for Disease Control and Prevention, 2014.

Brosi, Matthew, John D. Foubert, R. Sean Bannon, et al. "Effects of Women's Pornography Use on Bystander Intervention in a Sexual Assault Situation and Rape Myth Acceptance." *Oracle: The Research Journal of the Association of Fraternity/Sorority Advisors* 6, no. 2 (2011): 26–35.

Brown, Lyn Mikel, and Carol Gilligan. *Meeting at the Crossroads: Women's Psychology and Girls' Development.* New York: Ballantine Books, 1993.

Brown, Jane D. "Mass Media Influences on Sexuality." *Journal of Sex Research* 39, no. 1 (2002): 42–45.

Brown, Jane D., and Kelly L. L'Engle. "X-Rated: Sexual Attitudes and Behaviors Associated with U.S. Early Adolescents' Exposure to Sexually Explicit Media." *Communication Research* 36, no. 1 (2009): 129–51.

Brückner, Hannah, and Peter Bearman. "After the Promise: The STD Consequences of Adolescent Virginity Pledges." *Journal of Adolescent Health* 36 (2005): 271–78.

Brugman, Margaret, Sandra L. Caron, and Jany Rademakers. "Emerging Adolescent Sexuality: A Comparison of American and Dutch College Women's Experiences." *International Journal of Sexual Health* 22, no. 1 (2010): 32–46.

Brumberg, Joan Jacobs. *The Body Project: An Intimate History of American Girls.* New York: Random House, 1997.

Burns, April, Valerie A. Futch, and Deborah L. Tolman. "It's Like Doing Homework." *Sexuality Research and Social Policy* 7, no. 1 (2011).

Calogero, Rachel M. "Objects Don't Object: Evidence That Self-Objectification Disrupts Women's Social Activism." *Psychological Science* 24, no. 3 (2013): 312–18.

Cantor, David, Bonnie Fisher, Susan Chibnall, et al. *Report on the AAU Campus Climate Survey on Sexual Assault and Sexual Misconduct.* Washington, DC: Association of American Universities, 2015.

Carey, Kate, Sarah Durney, Robyn Shepardson, et al. "Incapacitated and Forcible Rape of College Women: Prevalence Across the First Year." *Journal of Adolescent Health* 56 (2015): 678–80.

Caron, Sandra L. *The Sex Lives of College Students: Two Decades of Attitudes and Behaviors.* Orono: Maine College Press, 2013.

Carpenter, Laura M. *Virginity Lost: An Intimate Portrait of First Sexual Experiences.* New York: New York University Press, 2005.

Carr, Joetta L., and Karen M. VanDeusen. "Risk Factors for Male Sexual Aggression on College Campuses." *Journal of Family Violence* 19, no. 5 (2004): 279–89.

Carroll, Jason S., et al. "Generation XXX: Pornography Acceptance and Use Among Emerging Adults." *Journal of Adolescent Research* 23 (2008): 6–30.

Centers for Disease Control and Prevention. "Binge Drinking: A Serious, Under-Recognized Problem Among Women and Girls." *CDC Vital Signs* (blog), January 2013. Atlanta: Centers for Disease Control and Prevention.

———. "Caffeine and Alcohol." *Fact Sheet*, November 19, 2014. Atlanta: Centers for Disease Control and Prevention.

———. "Reproductive Health: Teen Pregnancy, About Teen Pregnancy," 2014. Atlanta: Centers for Disease Control and Prevention.

———. "Youth Risk Behavior Surveillance." *Morbidity and Mortality Weekly Report*, June 13, 2014. Atlanta: Centers for Disease Control and Prevention.

Chambers, Wendy C. "Oral Sex: Varied Behaviors and Perceptions in a College Population." *Journal of Sex Research* 44, no. 1 (2007): 28–42.

Chandra, Anjani, William D. Mosher, Casey E. Copen, et al. "Sexual Behavior, Sexual Attraction, and Sexual Identity in the United States: Data from the 2006–2008 National Survey of Family Growth." *National Health Statistics Reports* 36, March 3, 2011. Washington, DC: U.S. Department of Health and Human Services.

Child Trends DataBank. "Oral Sex Behaviors Among Teens." Bethesda, MD: Child Trends DataBank, 2013.

Chyng, Sun, Ana Bridges, Robert Wosnitzer, et al. "A Comparison of Male and Female Directors in Popular Pornography: What Happens When Women Are at the Helm?" *Psychology of Women Quarterly* 32, no. 3 (2008): 312–25.

Collins, W. Andrew, Deborah P. Welsh, and Wyndol Furman. "Adolescent Romantic Relationships." *Annual Review of Psychology* 60 (2009): 631–52.

"Consent: Not Actually That Complicated," Rockstardinosaurpirateprincess.com, March 2, 2015.

Copen, Casey E., Anjani Chandra, and Gladys Martinez. "Prevalence and Timing of Oral Sex with Opposite-Sex Partners Among

Females and Males Aged 15–24 Years: United States 2007–2010," *National Health Statistics Reports* 56 (August 16, 2012).

Corinna, Heather. *S.E.X.: The All-You-Need-to-Know Progressive Sexuality Guide to Get You Through High School and College.* Boston: Da Capo Press, 2007.

Cornell, Jodi L., and Bonnie L. Halpern-Felsher. "Adolescent Health Brief: Adolescents Tell Us Why Teens Have Oral Sex." *Journal of Adolescent Health* 38 (2006): 299–301.

Daniels, Elizabeth A. "Sex Objects, Athletes, and Sexy Athletes: How Media Representations of Women Athletes Can Impact Adolescent Girls and Young Women." *Journal of Adolescent Research* 24 (2009): 399–422.

The Date Rape Backlash: Media and the Denial of Rape. Transcript. Documentary produced by Sut Jhally, 1994.

Davis, Kelly Cue. "The Influence of Alcohol Expectancies and Intoxication on Men's Aggressive Unprotected Sexual Intentions." *Experimental and Clinical Psychopharmacology* 18, no. 5 (2010): 418–28.

Diamond, Lisa. "Introduction: In Search of Good Sexual-Developmental Pathways for Adolescent Girls." In *Rethinking Positive Adolescent Female Sexual Development.* Edited by Lisa Diamond. San Francisco: Jossey-Bass, 2006, pp. 1–7.

Diamond, Lisa, and Ritch Savin-Williams. "Adolescent Sexuality." In *Handbook of Adolescent Psychology.* Edited by Richard M. Lerner and Laurence Steinberg. 3rd ed. New York: Wiley, 2009, pp. 479–523.

Dillard, Katie. "Adolescent Sexual Behavior: Demographics." 2002. Advocates for Youth, Washington, DC.

Dotson-Blake, Kylie P., David Knox, and Marty E. Zusman. "Exploring Social Sexual Scripts Related to Oral Sex: A Profile of College Student Perceptions." *Professional Counselor* 2 (2012): 1–11.

Douglass, Marcia, and Lisa Douglass. *Are We Having Fun Yet? The Intelligent Woman's Guide to Sex.* New York: Hyperion, 1997.

Dunn, Hailee, A. Gjelsvik, D. N. Pearlman, et al. "Association Between Sexual Behaviors, Bullying Victimization and Suicidal Ideation in a National Sample of High School Students:

Implications of a Sexual Double Standard." *Women's Health Issues* 24, no. 5 (2014): 567–74.

Edwards, Sarah R., Kathryn A. Bradshaw, and Verlin B. Hinsz. "Denying Rape but Endorsing Forceful Intercourse: Exploring Differences Among Responders." *Violence and Gender* 1, no. 4 (2014): 188–93.

England, Paula, et al. "Hooking Up and Forming Romantic Relationships on Today's College Campuses." In *Gendered Society Reader*. Edited by Michael S. Kimmel and Amy Aronson. 3rd ed. New York: Oxford University Press, 2008.

Englander, Elizabeth. "Low Risk Associated with Most Teenage Sexting: A Study of 617 18-Year-Olds." *MARC Research Reports*, Paper 6, 2012. Bridgewater, MA: Virtual Commons–Bridgewater State University.

Exner-Cortens, Deinera, John Eckenrode, and Emily Rothman. "Longitudinal Associations Between Teen Dating Violence Victimization and Adverse Health Outcomes." *Pediatrics* 131, no. 1 (2013): 71–78.

Fardouly, Jasmine, Phillipa C. Diedrichs, Lenny R. Vartanian, et al. "Social Comparisons on Social Media: The Impact of Facebook on Young Women's Body Image Concerns and Mood." *Body Image* 13 (2015): 38–45.

Fava, Nicole M., and Laina Y. Bay-Cheng. "Young Women's Adolescent Experiences of Oral Sex: Relation of Age of Initiation to Sexual Motivation, Sexual Coercion, and Psychological Functioning." *Journal of Adolescence* 30 (2012): 1–11.

Fay, Joe. "Teaching Teens About Sexual Pleasure." *SIEUS Report* 30, no. 4 (2002): 1–7.

Fine, Michelle. "Sexuality, Schooling, and Adolescent Females: The Missing Discourse of Desire." *Harvard Educational Review* 58 (1988): 29–53.

Fine, Michelle, and Sara McClelland. "Sexuality Education and Desire: Still Missing After All These Years." *Harvard Educational Review* 76 (2006): 297–338.

Finer, Lawrence B., and Jesse M. Philbin. "Sexual Initiation,

Contraceptive Use, and Pregnancy Among Young Adolescents." *Pediatrics* 131, no. 5 (2013): 886–91.

Finkelhor, David, Heather Turner, and Richard Ormrod. "Children's Exposure to Violence: A Comprehensive National Survey." *Juvenile Justice Bulletin*, October 2009.

Fisher, Deborah A., Douglas L. Hill, Joel W. Grube, et al. "Televised Sexual Content and Parental Mediation: Influences on Adolescent Sexuality." *Media Psychology* 12, no. 2 (2009): 121–47.

Fisher, Maryanne, Kerry Worth, Justin Garcia, et al. "Feelings of Regret Following Uncommitted Sexual Encounters in Canadian University Students." *Culture, Health and Sexuality* 14, no. 1 (2012): 45–57.

Fisher, William A. "All Together Now: An Integrated Approach to Preventing Adolescent Pregnancy and STD/HIV Infection." *SIECUS Report* 18, no. 4 (1990): 1–14.

Fortenberry, Dennis J. "Puberty and Adolescent Sexuality." *Hormones and Behavior* 64, no. 2 (2013): 280–87.

Fortenberry, Dennis J., Vanessa Schick, Debby Herbenick, et al. "Sexual Behaviors and Condom Use at Last Vaginal Intercourse: A National Sample of Adolescents Ages 14 to 17 Years." *Journal of Sexual Medicine* 7, suppl. 5 (2010): 305–14.

Foubert, John D., Matthew W. Brossi, and R. Sean Bannon. "Pornography Viewing Among Fraternity Men: Effects on Bystander Intervention, Rape Myth Acceptance, and Behavioral Intent to Commit Sexual Assault." *Sexual Addiction and Compulsivity: The Journal of Treatment and Prevention* 18, no. 4 (2011): 212–31.

Foubert, John D., Jonathan T. Newberry, and Jerry L. Tatum. "Behavior Differences Seven Months Later: Effects of a Rape Prevention Program on First-Year Men Who Join Fraternities." *NASPA Journal* 44 (2007): 728–49.

Ford, Jessie, and Paula England. "What Percent of College Women Are Sexually Assaulted in College?" Contexts.com, January 12, 2015.

Fox, Jesse, Rachel A. Ralston, Cody K. Cooper, et al. "Sexualized Avatars Lead to Women's Self-Objectification and Acceptance of Rape Myths." *Psychology of Women Quarterly*, October 2014.

Fredrickson, Barbara, et al. "That Swimsuit Becomes You: Sex Differences in Self-Objectification, Restrained Eating, and Math Performance." *Journal of Personality and Social Psychology* 75 (1998): 269–84.

Fredrickson, Barbara, and Tomi-Ann Roberts. "Objectification Theory: Toward Understanding Women's Lived Experience and Mental Health Risks." *Psychology of Women Quarterly* 21 (1997): 173–206.

Friedman, Jaclyn, and Jessica Valenti. *Yes Means Yes!: Visions of Female Sexual Power and a World Without Rape.* New York: Seal Press, 2008.

Garcia, Justin R., Chris Reiber, Sean G. Massey, et al. "Sexual Hook-Up Culture: A Review." *Review of General Psychology* 16, no. 2 (2012): 161–76.

Gerressu, Makeda, et al. "Prevalence of Masturbation and Associated Factors in a British National Probability Survey." *Archives of Sexual Behavior* 37 (2008): 266–78.

Gilligan, Carol, Nona Lyons, and Trudy Hanmer, eds. *Making Connections: The Relational Worlds of Adolescent Girls at Emma Willard School.* Cambridge, MA: Harvard University Press, 1990.

GLSEN. *Out Online: The Experiences of Lesbian, Gay, Bisexual and Transgender Youth on the Internet.* New York: GLSEN, July 10, 2013.

Gomillion Sarah C., and Traci A. Giuliano. "The Influence of Media Role Models on Gay, Lesbian, and Bisexual Identity." *Journal of Homosexuality* 58 (2011): 330–54.

Grello, Catherine M., et al. "No Strings Attached: The Nature of Casual Sex in College Students." *Journal of Sex Research* 43 (2006): 255–67.

Grunbaum, J. A., et al. "Youth Risk Behavior Surveillance— United States, 2001." *Morbidity and Mortality Weekly Report. CDC Surveillance Summaries* 51 (2002): 1–64.

Guttmacher Institute. "American Teens' Sexual and Reproductive Health." *Fact Sheet,* May 2014. New York: Guttmacher Institute.

———. "Sex and HIV Education." *State Policies in Brief,* June 1, 2015. New York: Guttmacher Institute.

Haffner, Debra W., ed. *Facing Facts: Sexual Health for America's*

Adolescents: The Report of the National Commission on Adolescent Sexual Health. Washington, DC: Sexuality Information and Education Council of the United States, 1995.

Halliwell, Emma, et al. "Are Contemporary Media Images Which Seem to Display Women as Sexually Empowered Actually Harmful to Women?" *Psychology of Women Quarterly* 35, no. 1 (2011): 38–45.

Halpern-Felsher, Bonnie L., Jodi L. Cornell, Rhonda Y. Kropp, and Jeanne M. Tschann. "Oral Versus Vaginal Sex Among Adolescents: Perceptions, Attitudes, and Behavior." *Pediatrics* 4 (2005): 845–51.

Hamilton, Laura, and Elizabeth A. Armstrong. "Gendered Sexuality in Young Adulthood: Double Binds and Flawed Options." *Gender and Society* 23 (2009): 589–616.

Harris, Michelle. "Shaved Paradise: A Sociological Study of Pubic Hair Removal Among Lehigh University Undergraduates." Senior thesis, 2009, Lehigh University, Bethlehem, PA.

Henry J. Kaiser Family Foundation. "Teen Sexual Activity." *Fact Sheet*, December 2002. Menlo Park, CA: Henry J. Kaiser Family Foundation.

Henry J. Kaiser Family Foundation/*YM* Magazine. *National Survey of Teens: Teens Talk About Dating, Intimacy, and Their Sexual Experiences.* Menlo Park, CA: Henry J. Kaiser Family Foundation, March 27, 1998.

Herbenick, Debby, et al. "Sexual Behavior in the United States: Results from a National Probability Sample of Men and Women Ages 14–94." *Journal of Sexual Medicine* 7, suppl. 5 (2010): 255–65.

Hirschman, Celeste, Emily A. Impett, and Deborah Schooler. "Dis/Embodied Voices: What Late-Adolescent Girls Can Teach Us About Objectification and Sexuality." *Sexuality Research and Social Policy* 3, no. 4 (2006): 8–20.

Hoff, Tina, Liberty Green, and Julia Davis. "National Survey of Adolescents and Young Adults: Sexual Health Knowledge, Attitudes and Experiences," 2004. Henry J. Kaiser Family Foundation, Menlo Park, CA, p. 14.

Horan, Patricia F., Jennifer Phillips, and Nancy E. Hagan. "The Meaning of Abstinence for College Students." *Journal of HIV/AIDS Prevention and Education for Adolescents and Children* 2, no. 2 (1998): 51–66.

Human Rights Campaign. *Growing Up LGBT in America.* Human Rights Campaign, Washington, DC, 2012.

Impett, Emily, Deborah Schooler, and Deborah Tolman. "To Be Seen and Not Heard: Femininity Ideology and Adolescent Girls' Sexual Health." *Archives of Sexual Behavior* 35 (2006): 129–42.

Impett, Emily, and Deborah Tolman. "Late Adolescent Girls' Sexual Experiences and Sexual Satisfaction." *Journal of Adolescent Research* 6 (2006): 628–46.

Joyner, Kara, and J. Richard Udry. "You Don't Bring Me Anything but Down: Adolescent Romance and Depression." *Journal of Health and Social Behavior* 41, no. 4 (2000): 369–91.

Kaestle, Christine Elizabeth. "Sexual Insistence and Disliked Sexual Activities in Young Adulthood: Differences by Gender and Relationship Characteristics." *Perspectives on Sexual and Reproductive Health* 41, no. 1 (2009): 33–39.

Kanin, Eugene J. "False Rape Allegations." *Archives of Sexual Behavior* 23, no. 1 (1994): 81–92.

Kann, Laura, Steven Kinchen, Shari L. Shanklin, et al. "Youth Risk Behavior Surveillance: United States, 2013." *Morbidity and Mortality Weekly Report.* Atlanta: Centers for Disease Control and Prevention, 2014.

Katz, Jennifer, and Vanessa Tirone. "Going Along with It: Sexually Coercive Partner Behavior Predicts Dating Women's Compliance with Unwanted Sex." *Violence Against Women* 16, no. 7 (2010): 730–42.

Kelly-Weeder, Susan. "Binge Drinking and Disordered Eating in College Students." *Journal of the American Academy of Nurse Practitioners* 23, no. 1 (2011): 33–41.

Kipnis, Laura. *The Female Thing: Dirt, Sex, Envy, Vulnerability.* New York: Pantheon Books, 2006.

Kirby, Douglas. *Emerging Answers 2007: Research Findings on Programs to Reduce Teen Pregnancy and Sexually Transmitted Diseases.* Washington, DC: National Campaign to Prevent Teen and Unplanned Pregnancy, 2007.

———. "Sex and HIV Programs: Their Impact on Sexual Behaviors of Young People Throughout the World." *Journal of Adolescent Health* 40 (2007): 206–17.

Kohler, Pamela K., Lisa E. Manhart, and William E. Lafferty. "Abstinence-Only and Comprehensive Sex Education and the Initiation of Sexual Activity and Teen Pregnancy." *Journal of Adolescent Health* 42 (2008): 334–51.

Krebs, Christopher P., Christine H. Lindquist, and Tara D. Warner. *The Campus Sexual Assault (CSA) Study Final Report*. Washington, DC: National Institute of Justice, 2007.

Kunkel, D., Keren Eyal, Keli Finnerty, et al. *Sex on TV 4*. Menlo Park, CA: Henry J. Kaiser Family Foundation, 2005.

Lamb, Sharon. "Feminist Ideals for a Healthy Female Adolescent Sexuality: A Critique." *Sex Roles* 62 (2010): 294–306.

Laumann, Edward O., Robert T. Michael, Gina Kolata, et al. *Sex in America: A Definitive Survey*. New York: Grand Central Publishing, 1995.

Lefkowitz, Bernard. *Our Guys: The Glen Ridge Rape and the Secret Life of the Perfect Suburb*. Berkeley: University of California Press, 1997.

Leigh, Barbara, and D. M. Morrison. "Alcohol Consumption and Sexual Risk-Taking in Adolescents." *Alcohol Health and Research World* 15 (1991): 58–63.

Lenhart, Amanda. "Teens and Sexting." Internet, Science, and Tech. Pew Research Center, December 15, 2009.

Lescano, Celia, et al. "Correlates of Heterosexual Anal Intercourse Among At-Risk Adolescents and Young Adults." *American Journal of Public Health* 99 (2009): 1131–36.

Levine, Judith. *Harmful to Minors: The Perils of Protecting Children from Sex*. Cambridge, MA: Da Capo Press, 2003.

Levy, Ariel. *Female Chauvinist Pigs: Women and the Rise of Raunch Culture*. New York: Free Press, 2006.

Lindberg, Laura Duberstein, Rachel Jones, and John S. Santelli. "Noncoital Sexual Activities Among Adolescents." *Journal of Adolescent Health* 43, no. 3 (2008): 231–38.

Lindberg, Laura Duberstein, John S. Santelli, and Susheela Singh. "Changes in Formal Sex Education: 1995–2002." *Perspectives on Sexual and Reproductive Health* 38 (2006): 182–89.

Lisak, David, Lori Gardinier, Sarah C. Nicksa, et al. "False Allegations of Sexual Assault: An Analysis of Ten Years of Reported Cases." *Violence Against Women* 16, no. 12 (2010): 1318–34.

Lisak, David, and Paul M. Miller Brown. "Repeat Rape and Multiple Offending Among Undetected Rapists." *Violence and Victims* 17, no. 1 (2002): 73–84.

Livingston, Jennifer, Laina Y. Bay-Cheng, et al. "Mixed Drinks and Mixed Messages: Adolescent Girls' Perspectives on Alcohol and Sexuality." *Psychology of Women Quarterly* 37, no. 1 (2013): 38–50.

Lounsbury, Kaitlin, Kimberly J. Mitchell, and David Finkelhor. "The True Prevalence of 'Sexting.'" *Fact Sheet*, April 2011. Crimes Against Children Research Center, Durham, NH.

Luker, Kristin. *When Sex Goes to School: Warring Views on Sex—and Sex Education—Since the Sixties.* New York: W. W. Norton, 2006.

Manago, Adriana, Michael B. Graham, Patricia M. Greenfield, et al. "Self-Presentation and Gender on MySpace." *Journal of Applied Developmental Psychology* 29, no. 6 (2008): 446–58.

Martinez, Gladys, Casey E. Copen, and Joyce C. Abma. "Teenagers in the United States: Sexual Activity, Contraceptive Use, and Childbearing, 2006–2010 National Survey of Family Growth." *Vital Health Statistics* 31 (2011): 1–35.

Martino, Steven C., Rebecca L. Collins, Marc N. Elliott, et al. "It's Better on TV: Does Television Set Teenagers Up for Regret Following Sexual Initiation?" *Perspectives on Sexual and Reproductive Health* 41, no. 2 (2009): 92–100.

McAnulty, Richard D., and Arnie Cann. "College Student Dating in Perspective: 'Hanging Out,' 'Hooking Up,' and Friendly Benefits." In *Sex in College.* Edited by Richard D. McAnulty. Santa Barbara, CA: Praeger, 2012, pp. 1–18.

McClelland, Sara I. "Intimate Justice: A Critical Analysis of Sexual Satisfaction." *Social and Personality Psychology Compass* 4, no. 9 (2010): 663–80.

———. "What Do You Mean When You Say That You Are Sexually Satisfied? A Mixed Methods Study." *Feminism and Psychology* 24, no. 1 (2014): 74–96.

————. "Who Is the 'Self' in Self-Reports of Sexual Satisfaction? Research and Policy Implications." *Sexuality Research and Social Policy* 8, no. 4 (2011): 304–20.

Meier, Evelyn P., and James Gray. "Facebook Photo Activity Associated with Body Image Disturbance in Adolescent Girls." *Cyberpsychology, Behavior, and Social Networking* 10, no. 10 (2013).

Mohler-Kuo, Meichun, George W. Dowdall, Mary P. Koss, et al. "Correlates of Rape While Intoxicated in a National Sample of College Women." *Journal of Studies on Alcohol* 65, no. 1 (2004).

Monk-Turner, Elizabeth, and H. Christine Purcell. "Sexual Violence in Pornography: How Prevalent Is It?" *Gender Issues* 2 (1999): 58–67.

Monroe, Scott M., Paul Rhode, John R. Seeley, et al. "Life Events and Depression in Adolescence: Relationship Loss as a Prospective Risk Factor for First Onset of Major Depressive Disorder." *Journal of Abnormal Psychology* 180, no. 4 (1999): 606–14.

Moore, Mignon R., and Jeanne Brooks-Gunn. "Healthy Sexual Development: Notes on Programs That Reduce Risk of Early Sexual Initiation and Adolescent Pregnancy." In *Reducing Adolescent Risk: Toward an Integrated Approach*. Edited by Daniel Romer. Thousand Oaks, CA: Sage Publications, 2003.

Moran, Caitlin. *How to Be a Woman*. New York: HarperPerennial, 2012.

Moran, Jeffrey. *Teaching Sex: The Shaping of Adolescence in the Twentieth Century*. Cambridge, MA: Harvard University Press, 2002.

National Center for Chronic Disease Prevention and Health Promotion. "Binge-Drinking: A Serious, Unrecognized Problem Among Women and Girls." *CDC Vital Signs* (blog), 2013. National Center for Chronic Disease Prevention and Health Promotion. Atlanta: Centers for Disease Control and Prevention, 2013.

National Institute on Alcohol Abuse and Alcoholism. "College Drinking." *Fact Sheet*, April 5, 2015. Washington, DC: National Institute on Alcohol Abuse and Alcoholism.

Norris, Jeanette, Kelly Cue Davis, William H. George, et al. "Alcohol's Direct and Indirect Effects on Men's Self-Reported Sexual

Aggression Likelihood." *Journal of Studies on Alcohol* 63 (2002): 688–69.

Norris, Jeanette, William H. George, Kelly Cue Davis, Joel Martell, and R. Jacob Leonesio. "Alcohol and Hypermasculinity as Determinants of Men's Empathic Responses to Violent Pornography." *Journal of Interpersonal Violence* 14 (1999): 683–700.

Orchowski, Lindsay M., Alan Berkowitz, Jesse Boggis, et al. "Bystander Intervention Among College Men: The Role of Alcohol and Correlates of Sexual Aggression." *Journal of Interpersonal Violence* (2015): 1–23.

Orenstein, Peggy. *Cinderella Ate My Daughter: Dispatches from the Front Lines of the New Girlie-Girl Culture.* New York: HarperPaperbacks, 2012.

———. *Flux: Women on Sex, Work, Love, Kids, and Life in a Half-Changed World.* New York: Anchor, 2001.

———. *Schoolgirls: Young Women, Self-Esteem, and the Confidence Gap.* New York: Anchor, 1995.

O'Sullivan, Lucia, et al. "I Wanna Hold Your Hand: The Progression of Social, Romantic and Sexual Events in Adolescent Relationships." *Perspectives in Sexual and Reproductive Health* 39, no. 2 (2007): 100–107.

Oswalt, Sara B., Kenzie A. Cameron, and Jeffrey Koob. "Sexual Regret in College Students." *Archives of Sexual Behavior* 34 (2005): 663–69.

Owen, Janice, G. K. Rhoades, S. M. Stanley, et al. "'Hooking up' Among College Students: Demographic and Psychosocial Correlates." *Archives of Sexual Behavior* 39 (2010): 653–63.

Paul, Elizabeth L., et al., "'Hookups': Characteristics and Correlates of College Students' Spontaneous and Anonymous Sexual Experiences." *Journal of Sexual Research* 37 (2000): 76–88.

Paul, Pamela. *Pornified: How Pornography Is Transforming Our Lives, Our Relationships, and Our Families.* New York: Times Books, 2005.

Peter, Jochen, and Patti Valkenburg. "Adolescents' Exposure to a Sexualized Media Environment and Notions of Women as Sex Objects." *Sex Roles* 56 (2007): 381–95.

———. "Adolescents' Exposure to Sexually Explicit Online Material and Recreational Attitudes Toward Sex." *Journal of Communication* 56, no. 4 (2006): 639–60.

————. "The Use of Sexually Explicit Internet Material and Its Antecedents: A Longitudinal Comparison of Adolescents and Adults." *Archives of Sexual Behavior* 40, no. 5 (October 2011): 1015–25.

Peterson, Zoe D., and Charlene L. Muehlenhard. "What Is Sex and Why Does It Matter? A Motivational Approach to Exploring Individuals' Definitions of Sex." *Journal of Sex Research* 44, no. 3 (2007): 256–68.

Phillips, Lynn M. *Flirting with Danger: Young Women's Reflections on Sexuality and Domination.* New York: New York University Press, 2000.

Pittman, Melissa. "The Joy of Slash: Why Do Women Want It?" *The High Hat*, Spring 2005.

Ponton, Lynn. *The Sex Lives of Teenagers: Revealing the Secret World of Adolescent Boys and Girls.* New York: Dutton, 2000.

Raphael, Jody. *Rape Is Rape: How Denial, Distortion, and Victim Blaming Are Fueling a Hidden Acquaintance Rape Crisis.* Chicago: Chicago Review Press, 2013.

Rector, Robert E., Kirk A. Jonson, and Laura R. Noyes. *Sexually Active Teenagers Are More Likely to Be Depressed and to Attempt Suicide: A Report of the Heritage Center for Data Analysis.* Washington, DC: Heritage Foundation, Center for Data Analysis, 2003.

Regnerus, Mark. *Forbidden Fruit: Sex and Religion in the Lives of American Teenagers.* New York: Oxford University Press, 2007.

———. "Porn Use and Support of Same-Sex Marriage." *Public Discourse*, December 20, 2012.

Remez, Lisa. "Oral Sex Among Adolescents: Is It Sex or Is It Abstinence?" *Family Planning Perspectives* 32 (2000): 298–304.

Ringrose, Jessica, Rosalind Gill, Sonia Livingstone, et al. *A Qualitative Study of Children, Young People, and 'Sexting': A Report Prepared for the NSPCC.* London: National Society for the Prevention of Cruelty to Children, 2012.

Robbins, Cynthia, Vanessa Schick, Michael Reece, et al. "Prevalence, Frequency, and Associations of Masturbation with Other Sexual

Behaviors Among Adolescents Living in the United States of America." *Archives of Pediatric and Adolescent Medicine* 165, no. 12 (2011): 1087–93.

Rosenbaum, Janet Elise. "Patient Teenagers? A Comparison of the Sexual Behavior of Virginity Pledgers and Matched Nonpledgers." *Pediatrics* 123 (2009): 110–20.

Ryan, Caitlin. "Generating a Revolution in Prevention, Wellness, and Care for LGBT Children and Youth." *Temple Political and Civil Rights Law Review* 23, no. 2 (2014): 331–44.

Sanders, Stephanie, Brandon J. Hill, William L. Yarber, et al. "Misclassification Bias: Diversity in Conceptualisations About Having 'Had Sex.'" *Sexual Health* 7, no. 1 (2010): 31–34.

Schalet, Amy T. "The New ABCD's of Talking About Sex with Teenagers." *Huffington Post*, November 2, 2011.

———. *Not Under My Roof: Parents, Teens, and the Culture of Sex.* Chicago: University of Chicago Press, 2011.

Schear, Kimberlee S. *Factors That Contribute to, and Constrain, Conversations Between Adolescent Females and Their Mothers About Sexual Matters.* Urbana, IL: Forum on Public Policy, 2006.

Schick, Vanessa R., Sarah K. Calabrese, Brandi N. Rima, et al. "Genital Appearance Dissatisfaction: Implications for Women's Genital Image Self-Consciousness, Sexual Esteem, Sexual Satisfaction, and Sexual Risk." *Psychology of Women Quarterly* 34 (2010): 394–404.

Sedgh, Gilda, Lawrence B. Finer, Akinrinola Bankole, et al. "Adolescent Pregnancy, Birth, and Abortion Rates Across Countries: Levels and Recent Trends." *Journal of Adolescent Health* 58, no. 2 (2012): 223–30.

Senn, Charlene Y., Misha Eliasziw, Paula C. Barata, et al. "Efficacy of a Sexual Assault Resistance Program for University Women." *New England Journal of Medicine* 372 (2015): 2326–35.

"Sexual Health of Adolescents and Young Adults in the United States." *Fact Sheet*, August 20, 2104. Menlo Park, CA: Henry J. Kaiser Family Foundation.

Sharpley-Whiting, Tracy D. *Pimps Up, Ho's Down: Hip Hop's Hold on Young Black Women.* New York: New York University Press, 2008.

SIECUS. "A Brief History of Federal Funding for Sex Education and Related Programs." *Fact Sheet*. Washington, DC: SIECUS, n.d.

———. "Questions and Answers: Adolescent Sexuality." Washington, DC: SIECUS, November 12, 2012.

———. "What Research Says . . . Comprehensive Sex Education." *Fact Sheet*. Washington, DC: SIECUS, October 2009.

Simmons, Rachel. *The Curse of the Good Girl: Raising Authentic Girls with Courage and Confidence*. Reprint. New York: Penguin, 2010.

———. *Odd Girl Out, Revised and Updated: The Hidden Culture of Aggression in Girls*. New York: Mariner Books, 2011.

Simpson, Jeffry A., W. Andrew Collins, and Jessica E. Salvatore. "The Impact of Early Interpersonal Experience on Adult Romantic Relationship Functioning: Recent Findings from the Minnesota Longitudinal Study of Risk and Adaptation." *Current Directions in Psychological Science* 20, no. 6 (2011): 355–59.

Sinozich, Sofi, and Lynn Langton. *Special Report: Rape and Sexual Assault Victimization Among College-Age Females, 1995–2013*. Washington, DC: Office of Justice Programs, Bureau of Justice Statistics, U.S. Department of Justice, 2014.

Slater, Amy, and Marika Tiggeman. "A Test of Objectification Theory in Adolescent Girls." *Sex Roles* 46, no. 9/10 (May 2002): 343–49.

Sommers, Christina Hoff. *Who Stole Feminism? How Women Have Betrayed Women*. New York: Simon & Schuster, 1994.

Steering Committee on Undergraduate Women's Leadership at Princeton University. *Report of the Steering Committee on Undergraduate Women's Leadership*. Princeton, NJ: Princeton University, 2011.

Stermer, S. Paul, and Melissa Burkley. "SeX-Box: Exposure to Sexist Video Games Predicts Benevolent Sexism." *Psychology of Popular Media Culture* 4, no. 1 (2015): 47–55.

Steyer, James. *Talking Back to Facebook: The Common Sense Guide to Raising Kids in the Digital Age*. New York: Scribner, 2012, pp. 22–23.

Strasburger, Victor. "Policy Statement from the American Academy of Pediatrics: Sexuality, Contraception, and the Media." *Pediatrics* 126, no. 3 (September 1, 2010): 576–82.

Tanenbaum, Leora. *Slut: Growing Up Female with a Bad Reputation.* New York: HarperPerennial, 2002.

Thomas, J. "Virginity Pledgers Are Just as Likely as Matched Nonpledgers to Report Premarital Intercourse." *Perspectives on Sexual and Reproductive Health* 41, no. 63 (March 2009).

Thompson, Sharon. *Going All the Way: Teenage Girls' Tales of Sex, Romance, and Pregnancy.* New York: Hill and Wang, 1995.

Tolman, Deborah. *Dilemmas of Desire: Teenage Girls Talk About Sexuality.* Cambridge, MA: Harvard University Press, 2002.

Tolman, Deborah, Brian R. Davis, and Christin P. Bowman. "That's Just How It Is": A Gendered Analysis of Masculinity and Femininity Ideologies in Adolescent Girls' and Boys' Heterosexual Relationships." *Journal of Adolescent Research* (June 2015).

Tolman, Deborah, Emily Impett, et al. "Looking Good, Sounding Good: Femininity Ideology and Adolescent Girls' Mental Health." *Psychology of Women Quarterly* 30 (2006): 85–95.

Trenholm, Christopher, Barbara Devaney, Ken Fortson, et al. "Impacts of Four Title V, Section 510 Abstinence Education Programs." Office of the Assistant Secretary for Planning and Evaluation, U.S. Department of Health and Human Services. Princeton, NJ: Mathematics Policy Research, 2007.

U.S. House of Representatives, Committee on Government Reform Minority Staff, Special Investigations Division. *The Content of Federally Funded Abstinence-Only Education Programs.* Prepared for Rep. Henry A. Waxman. Washington, DC: U.S. Government Printing Office, 2004.

U.S. Senate Subcommittee on Financial and Contracting Oversight. *Sexual Violence on Campus.* By Claire McCaskill. 113th Congress. Senate Report, July 9, 2014.

Valenti, Jessica. *The Purity Myth: How America's Obsession with Virginity Is Hurting Young Women.* Berkeley, CA: Seal Press, 2009.

Vanwesenbeeck, Ine. "Sexual Health Behaviour Among Young People in the Netherlands." Presentation at the Sexual Health Forum, Brussels, March 13, 2009.

Vernacchio, Al. *For Goodness Sex: Changing the Way We Talk to Teens About Sexuality, Values, and Health.* New York: HarperWave, 2014.

Wade, Lisa, and Caroline Heldman. "Hooking Up and Opting Out." In *Sex for Life: From Virginity to Viagra, How Sexuality Changes Throughout Our Lives.* Edited by Laura Carpenter and John DeLamater. New York: New York University Press, 2012, pp. 128–45.

Ward, L. Monique. "Understanding the Role of the Entertainment Media in the Sexual Socialization of American Youth: A Review of Empirical Research." *Developmental Review* 23 (2003): 347–88.

Ward, L. Monique, Edwina Hansbrough, and Eboni Walker. "Contributions of Music Video Exposure to Black Adolescents' Gender and Sexual Schemas." *Journal of Adolescent Research* 20 (2005): 143–66.

Ward, L. Monique, and Kimberly Friedman. "Using TV as a Guide: Associations Between Television Viewing and Adolescents' Sexual Attitudes and Behavior." *Journal of Research on Adolescence* 16 (2006): 133–56.

Widerman, Michael M. "Women's Body Image Self-Consciousness During Physical Intimacy with a Partner." *Journal of Sex Research* 37, no. 1 (2000): 60–68.

Widman, Laura, et al. "Sexual Communication and Contraceptive Use in Adolescent Dating Couples." *Journal of Adolescent Health* 39 (2006): 893–99.

Wolak, Janis, Kimberly Mitchell, and David Finkelhor. "Unwanted and Wanted Exposure to Online Pornography in a National Sample of Youth Internet Users." *Pediatrics* 119, no. 2 (2007): 247–57.

Wright, Paul J. "Show Me the Data! Empirical Support for the 'Centerfold Syndrome.'" *Psychology of Men and Masculinity* 13, no. 2 (2011): 180–98.

———. "A Three-Wave Longitudinal Analysis of Preexisting Beliefs, Exposure to Pornography, and Attitude Change." *Communication Reports* 26, no. 1 (2013): 13–25.

Wright, Paul J., and Michelle Funk. "Pornography Consumption and Opposition to Affirmative Action for Women: A Prospective Study." *Psychology of Women Quarterly* 38, no. 2 (2013): 208–21.

Wright, Paul J., and Robert S. Tokunaga. "Activating the Centerfold Syndrome: Recency of Exposure, Sexual Explicitness, Past Exposure to Objectifying Media." *Communications Research* 20, no. 10 (2013): 1–34.

Yung, Corey Rayburn. "Concealing Campus Sexual Assault: An Empirical Examination." *Psychology, Public Policy, and Law* 21, no. 1 (2015): 1–9.

Index

About the Author

Peggy Orenstein is the *New York Times* bestselling author of *Cinderella Ate My Daughter*, *Waiting for Daisy*, *Flux*, and *Schoolgirls*. A contributing writer for the *New York Times Magazine*, she has been published in *USA Today*, *Parenting*, *Salon*, *The New Yorker*, and other publications, and has contributed commentary to NPR's *All Things Considered*. She lives in Northern California with her husband and daughter.

ALSO BY PEGGY ORENSTEIN

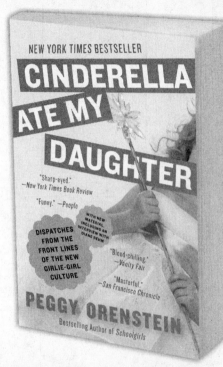

CINDERELLA ATE MY DAUGHTER
Dispatches from the Front Lines of the
New Girlie-Girl Culture

Available in Paperback, E-book, and Digital Audio

**"This addictively readable book manages, somehow,
to be simultaneously warm and chilling."**
—Rebecca Traister, author of *Big Girls Don't Cry*

The rise of the girlie-girl, warns Peggy Orenstein, is no innocent phenomenon. Following her acclaimed books *Flux, Schoolgirls*, and the provocative *New York Times* bestseller *Waiting for Daisy, Cinderella Ate My Daughter* offers a radical, timely wake-up call for parents, revealing the dark side of a pretty and pink culture confronting girls at every turn as they grow into adults.